Biochemistry and Molecular Biology in the Post Genomic Era

nova
science publishers
www.novapublishers.com

Biochemistry Research Trends

Water in Biology: A Molecular View
Michael E. Green, PhD (Editor)
Alisher M Kariev (Editor)
2023. ISBN: 979-8-88697-708-0 (Hardcover)
2023. ISBN: 979-8-88697-754-7 (eBook)

Biomolecules and Corrosion
Santosh Kumar Karn, PhD (Editor)
Anne Bhambri (Editor)
2023. ISBN: 979-8-88697-458-4 (Softcover)
2023. ISBN: 979-8-88697-531-4 (eBook)

The Biochemical Guide to Enzymes
David Aebisher, PhD (Editor)
Dorota Bartusik-Aebisher, PhD (Editor)
2022. ISBN: 979-8-88697-410-2 (Softcover)
2022. ISBN: 979-8-88697-518-5 (eBook)

Mineral Water: From Basic Research to Clinical Applications
Maria João Martins, PhD (Editor)
2022. ISBN: 978-1-68507-458-6 (Hardcover)
2022. ISBN: 978-1-68507-541-5 (eBook)

Terpenes and Terpenoids: Sources, Applications and Biological Significance
Charles A. Davies (Editor)
2022. ISBN: 978-1-68507-559-0 (Hardcover)
2022. ISBN: 978-1-68507-595-8 (eBook)

Circadian Rhythms and Their Importance
Rajeshwar P. Sinha, PhD (Editor)
2022. ISBN: 978-1-68507-547-7 (Hardcover)
2022. ISBN: 978-1-68507-585-9 (eBook)

More information about this series can be found at
https://novapublishers.com/product-category/series/biochemistry-research-trends/

Rushika Patel
Editor

Biosensing

Methods, Applications and Technology

nova
science publishers
www.novapublishers.com

https://doi.org/10.52305/AFUD0235

NOTICE TO THE READER

Library of Congress Cataloging-in-Publication Data

Names: Patel, Rushika, editor.
Title: Biosensin: methods, applications and technology / Rushika Patel, PhD (editor), Research Associate, Gujarat Biotechnology Research Centre,
Department of Science, and Technology, Government of Gujarat, India and
Governing Council member of Wildlife & Conservation Biology Research Foundation, India.
Identifiers: LCCN 2023038890 (print) | LCCN 2023038891 (ebook) | ISBN 9798886979114 (paperback) | ISBN 9798891131354 (adobe pdf)
Subjects: LCSH: Biosensors.
Classification: LCC R857.B54 B53387 2023 (print) | LCC R857.B54 (ebook) | DDC 610.28--dc23/eng/20230824
LC record available at https://lccn.loc.gov/2023038890
LC ebook record available at https://lccn.loc.gov/2023038891

Published by Nova Science Publishers, Inc. † New York

To my parents Bharatkumar and Anjana, who have
taught me the meaning of love and sacrifice,
and to my husband Kunal Patel, who has been my
constant companion, supporter,
and my best friend. This book is dedicated to you both,
with all my love and appreciation.

Contents

Preface

The book *Biosensing: Methods, Application and Technology* is a comprehensive guide to the rapidly advancing field of biosensing. It covers the latest developments and trends in biosensor technology, and provides a detailed overview of the different methods, applications and technologies used in biosensing. The book is divided into several chapters, each of which focuses on a specific aspect of biosensing.

The book starts with an introduction to biosensors and their applications, including an overview of the different types of biosensors and their working principles. It then goes on to explore the various biosensing techniques and technologies currently in use, such as electrochemical, optical and microbial biosensors.

The book also covers the latest advances in biosensing, such as the use of nanomaterials in biosensors, and the development of sensors for specific applications, such as food safety, environmental monitoring and medical diagnostics. The book also provides an in-depth look at the challenges facing biosensing technology, and discusses the future direction of biosensing research and development.

The book is written by a group of experts in the field, providing readers with a wealth of knowledge and experience, and the book is expected to be a valuable resource for anyone interested in learning more about the latest developments and future directions in biosensing technology. The book is divided into five chapters, each of which delves into a specific aspect of biosensing.

Chapter 1 explores the various display technologies used in biosensors, providing a detailed overview of the different types of biosensors and their applications.

Chapter 2 delves into the topic of sensing environmental contaminants, discussing the current trends and future aspects of microbial-derived biosensors.

Chapter 3 covers the recent trends in the detection of staphylococcal enterotoxins in food matrices, providing an in-depth look at the challenges and opportunities in this area of research.

Chapter 4 focuses on nanobiosensors and their potential to revolutionize sustainable agriculture, providing an overview of the latest developments in this field.

Finally, Chapter 5 looks to the future of biosensing technology, discussing the potential of geno-sensors for sustainable future.

This book is an essential resource for researchers, scientists and engineers working in the field of biosensing, as well as for students and professionals interested in learning more about this exciting and rapidly-evolving field. It provides a detailed and up-to-date overview of the latest biosensing techniques and technologies, and offers valuable insights into the challenges and opportunities facing biosensing in the future.

Acknowledgments

Biosensing: Methods, Applications, and Technology is a comprehensive guide to the rapidly expanding field of biosensors. It is my honor to acknowledge my PhD supervisor Dr. Nasreen Munshi Nirma University and Research committee members Dr. Naresh Kumar, Head Biochemistry Department M.S. University; Dr. Haresh Keharia, Bioscience Department S.P. University Prof. Gerald Thouand, University of Nantes; Prof. VanDer Meer, University of Lausanne; Prof. Eric D Van Hullebusch, Université Paris Cité for their guidance and encouragement throughout my research journey in the field of biosensors. I am also grateful to anonymous reviewers for valuable insights and all the authors who have contributed valuable chapters to this book. My work on this project would not have been possible without the support of the current director of Gujarat Biotechnology Research Centre, Professor Chaitnya Joshi, and Joint Director of Gujarat Biotechnology Research Centre, Dr. Madhvi Joshi.

I would like to express my deep appreciation to Nova Science Publishers for their support and guidance throughout the development of this book. I am particularly grateful to Editorial Assistant, Ashley Holady whose thoughtful comments and suggestions greatly improved the manuscript. I also want to thank Marketing Team and Mr. Thomas for their hard work in getting the book to the public.

This book is a culmination of my passion for biosensor and the dedication of my parents and husband. Their unwavering support and encouragement helped me to not just complete but excel in my research and writing of this book. Their belief in me and my abilities pushed me to strive for excellence in every aspect of my academic and personal life. I am also thankful to my parents-in-law (Hema and Bhupen), sister Vipra, brother (Jeel), my small bundle of joy niece-Aadhya for the constant support and encouragement throughout my research and writing of this book. Their love and support have been the wind beneath my wings. It is because of their support that I can present this book as a comprehensive guide to the latest developments in biosensing. It is my hope that this book will be a valuable resource for scientists, engineers, researchers, and students working in the field of biosensors.

Chapter 1

Biosensors Based on Display Technologies

Nidhi Patil
Asheemita Bagchi
and Preeti Srivastava*
Department of Biochemical Engineering and Biotechnology
Indian Institute of Technology Delhi, Hauz Khas, New Delhi, India

Abstract

Biosensors utilize biological macromolecules to detect analytes. Biosensors are gaining a lot of interest for their high selectivity and sensitivity. Here, we describe the various *in vitro* and *in vivo* display technologies and their use for the construction of biosensors. The advantages and applications of the biosensors-based on display technologies will be discussed.

Keywords: phage display, yeast cell surface display, bacterial cell surface display, aptamers, DNA-DNA hybridization

1. Introduction

Biosensors are increasingly becoming popular over the past 25 years. The first ever biosensor created was in 1962, when Leland C Clark developed the enzyme electrodes. Biosensors are essentially small devices that have the ability to sense an environmental change in their vicinity by employing a

* Corresponding Author's Email: preeti@dbeb.iitd.ac.in; preetisrivastava@hotmail.com.

In: Biosensing
Editor: Rushika Patel
ISBN: 979-8-88697-911-4

biological, or physiological recognition element and also to convert that alteration or change to an electrical signal that usually is proportional to the analyte's concentration or to the target that is being monitored (Kavita V, 2017; Koopaee et al. 2020; Pancrazio et al. 1999).

2. Characteristics of Biosensors

Biosensors are characterized by the following properties: a) they are known to be selective and sometimes specific to biological analytes, b) they have the ability to detect the analytes and also convert that information into an electrical signal (Kavita, 2017; Lee et al. 2014; Pividori et al. 2000; Rashid and Yusof, 2017; Singh et al. 2012). An ideal biosensor must be portable, automated and must deliver the results in short durations. Another important aspect of a biosensor is linearity in the results which is representative of the accuracy of the device. It should be able to not only detect the analyte, but also measure a slight change in its concentration over a wide working range (Koopaee et al. 2020). The main advantage of biosensors is that they are inexpensive, and deliver rapid, sensitive results that do not require efficient and highly skilled technical personnel (Pancrazio et al. 1999).

Some problems associated with the extensive usage of biosensor technology are the stability and the lack of reproducibility of the results (Pancrazio et al. 1999).

3. Parts of a Biosensor

A biosensor consists of five parts traditionally, (i) an analyte or target that needs to be recognized or detected, (ii) a bioreceptor, having the ability to recognize that analyte, which detects the change (iii) transducer, a significant component as it has the responsibility for converting one form of energy to another, it converts the reaction between the analyte and the bioreceptor to a detectable signal (the process being called signalisation); they produce the output signal in optical or electrical forms that are usually proportional to the original analyte concentration, (iv) electronic circuitry associated with the process, amplifying the signal, converting it from an analogue to a digital form and presenting the signal in a display, and (v) a display for easy understanding of the results (Koopaee et al. 2020). Figure 1 is a schematic representation of the parts and their respective functions of a typical biosensor.

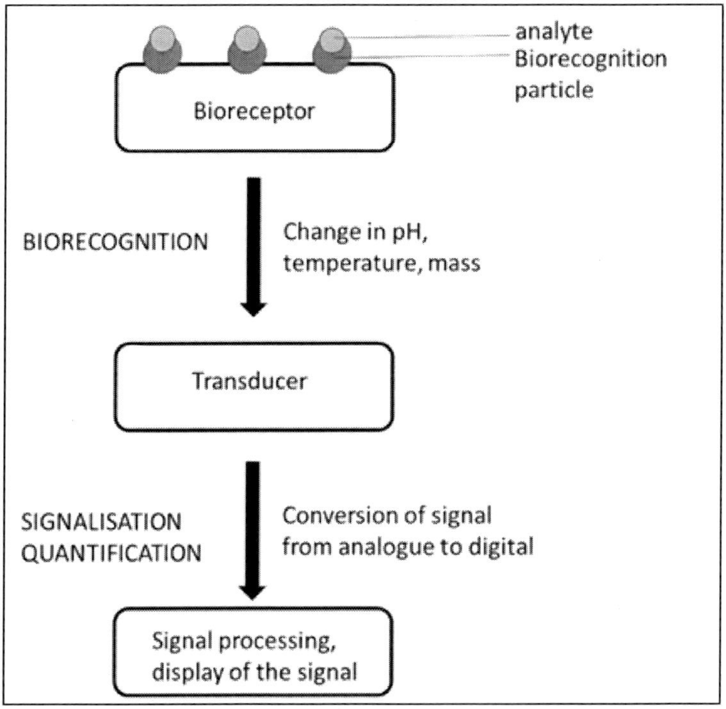

Figure 1. The basic structure of a biosensor, having a bioreceptor, transducer and a signal processor. The biochemical reaction which takes place when the analyte interacts with the immobilized ligand is sensed by the transducer which in turn converts the particular event to a readable signal (Pividori et al. 2000).

There are many types of biosensors, essentially based on the principle used by the transducer, for the parameters of measurement of the signal (Kavita V, 2017). They are:

1. *Amperometric:* wherein the current is measured at constant potential (Ho et al. 2004). It was reported that amperometry-based detection device would be the best possible candidate for the detection of morphine in blood and urine of patients undergoing surgical procedures to prevent overdose due to it being economical, sensitive and portable.

2. *Potentiometric:* measurement of voltage at constant current was used by Wang et al. 2001 for developing gold nanoparticles based on potentiometric detection of DNA hybridization (Wang et al. 2001). Potentiometric assays depend on measurement of variation of potential or pH, so the output signal that is generated is because of changes in concentrations of ionic species.
3. *Piezoelectric:* The development, uses and current knowledge of piezoelectric quartz crystals (PQC) in biosensors was reviewed (Bunde et al. 1998). These are types of acoustic sensors where the binding between the analyte or target and the biorecognition particle that has been adsorbed on the transducer surface brings about a mass change which is read and processed by the PQC. These biosensors produce real-time output which is also accompanied with the simplicity of usage and cost-effectiveness (Ivnitski et al. 1999).
4. *Thermal:* They have the ability to detect the release of heat energy accompanied with a biochemical reaction (Vasuki et al. 2019). They have found wide applicability in recent times for their simplicity, operational ease, cost effectiveness and stability.
5. *Optical:* They work by sensing a change in refractive index or thickness when the analyte binds to the bio-recognition particle on the surface of the transducer, these biosensors are widely used for their ease of operation during label-free detection of analytes. They offer many advantages, namely portability due to low cost, miniaturization and no interference of electrical signal (Mehrvar et al. 2000).

The bioreceptor can vary from tissue, microorganisms, whole cell, organelles, nucleic acids, proteins, enzymes, antibodies, etc. It is the bioreceptors whose nature determines the overall functionality of the biosensor. Here, we describe the biosensors constructed based on *in vivo* and *in vitro* technologies.

4. *In Vivo* Display Technologies

They utilize whole cells and the peptide of interest is made to display on the outer surface of cell using anchoring proteins.

4.1. Biosensors Based on Phage Display

A method of display in which exogenous polypeptides are expressed by phage particles and presented for binding to various target molecules is called as phage display. On cell surface of a filamentous phage, large peptides or protein libraries are displayed via which specific selection of proteins, peptides and antibodies with any target can be achieved. This technology was first demonstrated in 1986 by George Smith. In this technique, the gene encoding one of the viral coat proteins is cloned in frame with the gene encoding the desired peptide, creating a translational fusion. Thus, the desired peptide gets fused with one of the coat proteins and is assembled on the viral particle. Thus, a physical linkage among the viral particle and the displayed protein is achieved, which can further be manipulated for use in combinatorial peptide libraries in order to select a vast number of analytes and molecules of target (Figure 2). Table 1 lists phage display-based biosensors.

Table 1. Biosensors based on phage display

S. No	Organism/ Protein	Biosensor	Application	Reference
1	M13	SPR	Detection of *Listeria monocytogenes*	(Nanduri et al. 2007)
2	M13	SPR	Detection of pathogen – *Salmonella*	(Karoonuthaisiri et al. 2014)
3	M13-SWNT	SPR	Detection of biomarker	(Lee et al. 2016)
4	M13	SPR	Detection of pathogens - *E. coli*, *P. aeruginosa*, *V. cholerae*, *Xanthomonas campestris*	(Peng H et al. 2018))

One important benefit of using phage display technology is that phage particles are found to withstand very harsh conditions such as low temperatures and low pH without losing their bacterial infectivity.

Real-time monitoring of specific interactions between the peptide library and its target molecule has been achieved by using a flow system in biosensors which depend on the binding affinities depicted by their dissociation rates (Bradkovic T, 2010; Takakusagi et al. 2010).

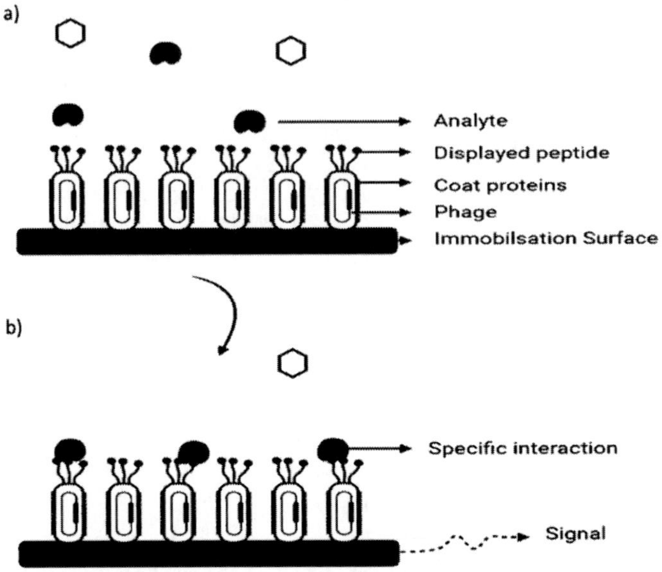

Figure 2. Phage cell surface display-based biosensors: a). Phage particles displaying peptide through pIII protein and immobilised on an inert surface. B). Binding of analyte and displayed protein leads to production of the signal.

4.1.1. Applications of Phage Display as Biosensors

A *Salmonella* specific binder bacteriophage having the peptide displayed on its surface was immobilised on a chip-based biosensor made using surface plasmon resonance technique. The biosensor measures the change in refractive index caused due to interaction between the immobilised bacteriophage and *Salmonella* leading to the detection of the pathogen. Several bacterial pathogen species have been identified by observing changes in surface plasmon resonance when interaction occurs between the bacterial species and its receptor displayed on bacteriophage.

An M13 bacteriophage was chimerically modified to display Receptor Binding Proteins of foreign phage capable of detecting specific pathogenic bacteria. The phage was then chemically modified to interact with gold nanoparticles which would react with target pathogen and facilitate SPR detection (Peng H and Chen IA, 2018).

M13 phage particle has been used wholly by fabricating on a carbon nanotube based nanomesh to develop an enzyme-based electrochemical biosensor. By biopanning from a pVIII peptide library, the phages which were able to bind to single walled carbon nanotube (SWNT) were selected and a nanomesh was assembled using both SWNT and the phage particles. A flexible microarray was incorporated with this nano mesh. This microarray enabled the detection of high-density signals. An enzyme-dependent biosensor making use of direct electron transfer (DET) detection was developed in combination with phage displayed CNT nanomesh. This biosensor successfully detected molecules such as glucose, lactate, peroxide, cholesterol, catechol, galactose, etc. by changing the immobilised enzymes like glucose oxidase, cholesterol oxidase, etc. (Lee SW et al. 2016).

Biosensors have been developed in which the pH insensitive Cy7 and the pH responsive HCy-646 have been modified and fused to be displayed on the phage. This gives rise to construction of a pH phage probe which is dependent on near infrared fluorescence. Many diseases disrupt the acid/base homeostasis of the body. Thus *in vivo* detection of such a change in pH in biological processes becomes essential. Fluorescence based imaging methodologies are preferred for investigating pH changes in body due to their high sensitivity and non-invasive nature. Several copies of NIR fluorophores have been functionalised on the M13 bacteriophage for targeted imaging application with receptor(s). The pVIII protein has been used due to high copy number and exposed amino terminals in the solvent which are then available for bioconjugation (Hilderbrand SA, 2008). Certain reporter enzymes along with target specific antibody have been functionalised on bacteriophages to develop immunochromatographic assays. Horseradish peroxidase (HRP) and analyte specific antibody fusion has also been displayed on the M13 bacteriophage. Immunochromatographic lateral flow assays are inexpensive diagnostic method used in preliminary testing of diverse biological samples. It is a qualitative as well as semi quantitative technique used for the detection of wide range of biomolecules like nucleic acids, toxins, proteins, hormones, bacterial and viral pathogens. Conventionally, nanoparticles made of gold, carbon, selenium have been coated with antibodies to detect the target molecule. Functionalised viral nanoparticles, developed using phage display technologies, provide a stronger signal per individual affinity agent and hence are used as scaffolds for chemically attaching multiple reporter enzymes (Adhikari M, 2013).

Another example where whole viral particles have been immobilised on a QCM biosensor surface has been demonstrated by Nanduri et al. They utilised phage particles as molecular recognition elements to be immobilised by simple physical adsorption. A phage particle displaying a single chain variable fragment having a specific affinity towards *Listeria monocytogenes* has been immobilised on the surface of a SPR to develop a biosensor (Nanduri et al. 2007).

4.2. Biosensors Based on Bacterial Cell Surface Display

In phage display system, the displayed foreign protein's size faced limitations. In microbial cell surface display, the heterologous protein's expression is carried out by fusion of the same with an anchoring motif which are the natural proteins of the cell surface. By the technique of cell surface display, the molecule displayed becomes freely accessible for binding or activity studies without any need for the target to pass through the membrane barrier (Jose J, 2006). The use of bacterial cell surface display system provides an advantage over the bacteriophage-based system as the bacteria are self-replicative and the size of the bacterial cell is sufficient enough for being analysed by optical visualizing methods such as microscopy or FACS (Francisco et al. 1993). Bacterial cell surface display-based biosensors are shown in Table 2. The first report of using bacterial cell surface display came in 1986 by Freud et al. and also Charbit et al. The proteins which need to be displayed on the surface of the bacterial cell are called as a passenger or target protein and the natural protein of the cell to which the protein is anchored is called as the carrier protein. Fusing the passenger protein with the carrier protein is translational in nature and the fusion is such that there is no loss in membrane integrity, no growth defect in the cell as well as maintenance of the cell wall binding property of the surface protein (Freudl et al. 1986; Charbit et al. 1986).

The primary requirement of the carrier protein is that it should have an anchoring domain which keeps the fusion protein on the cell surface without being detached and a strong signal peptide for transport through the inner membrane. It should be resistant to the action of proteases along with being compatible with the passenger protein sequences which are to be inserted for fusion. The examples of some widely used carrier proteins are bacterial fimbriae, ice nucleation proteins, S-layer proteins along many other outer membrane proteins. The location where the insertion of peptide is carried out is crucial as it determines the specific activity, stability, immobilization efficiency and post translational modifications.

Another important factor is the selection of a host for bacterial cell surface display. The host should possess compatibility with the protein that needs to be displayed and which is easy to be cultivated without lysing the cell. The host cell membrane should also have lower activities of proteases that are associated extracellularly to the cell wall. Due to fragile cell membranes of Gram-negative bacteria, for many cell surface display applications, bacteria which are Gram-positive are preferred. However, due to high transformation efficiency and ease of handling of *E. coli*, it is many times preferred. In order to make whole cell adsorbents and whole cell catalysts, Gram-positive bacteria are used owing to the rigid cell wall structure that they possess. *Bacillus* and *Staphylococcus* strains are most commonly used.

Depending upon the nature of the passenger and carrier proteins, the fusion strategy can vary from being a C-terminal fusion, an N-terminal fusion or a sandwich fusion.

Table 2. Biosensors based on bacterial cell surface display

S. No	Organism/Peptide	Biosensor	Application	Reference
1	Streptavidin on *E. coli*	SPR	Streptavidin-biotin interaction	(Park et al. 2009)
2	Metal binding peptide on *E. coli*	Fluorescence	Heavy metal detection	(Ravikumar et al. 2012)
3	Xylose dehydrogenase on *E. coli*	Electrochemical	Carbohydrate monitoring	(Li et al. 2013)
4	Organophosphorus hydrolase on *E. coli*	Fluorescence	Environmental monitoring of pesticides	(Liu et al. 2013)
5	Glutamate dehydrogenase on *E. coli*	Amperometric	L-Glutamate detection in Biomedicine, Food processing and Bioprocess monitoring.	(Liang et al. 2015)
6	Glucose oxidase on *E. coli*	Amperometric	Glucose detection	(Liang et al. 2015)
7	Laccase on *E. coli*	Electrochemical	Detection of adulterant	(Zhang et al. 2018)
8	Z domains *on E. coli*	SPR	Biomarker detection	(Jeon et al. 2018)
9	Cytokinin oxidase on *E. coli*	Electrochemical	Detecting hormones	(Li et al. 2019)
10	Acetaldehyde dehydrogenase on *E. coli*	Optical	Detection of carcinogen	(Liang et al. 2021)

Biosensors based on whole cells provide an advantage over their other counterparts due to the lesser cost involved with its development and production in high amounts, along with the improved stability provided by the whole cells. The production of high cell numbers is easy, and its cultivation is possible directly in large reactors. The growth conditions suitable for the micro-organisms can be easily monitored and necessary nutrients can be provided into the culture. The microorganisms are easy to manipulate and can withstand harsh conditions. The live organisms are coupled with different sensing techniques which can be either electrical or optical based. Any electroactive change produced on sensing can be detected by amperometer, impedimeter, conductimeter or potentiometer whereas the optical detection can be done based on bioluminescence, fluorescence or colorimetry.

A biosensor having the enzyme glucose dehydrogenase displayed on *E. coli* cell surface has been developed for detecting glucose levels. The INP protein was used as a motif to anchor for the fusion. To develop the biosensor, a multiwalled carbon nanotube was used and the signal detection was carried out amperometrically (Liang et al. 2013).

The amino acid L-glutamate is a crucial amino acid playing a role in the central nervous system as a neurotransmitter. In many organisms it serves the role of nitrogen and energy source. It is also used as a flavour enhancer in food industry. An excessive L-glutamate amount can cause neuro related problems like headache, stroke or even Parkinson's disease and also gastro related problems like stomach ache and indigestion. Thus, detection of L-glutamate in biological as well as food samples is necessary. Biosensors have been prepared to make use of Glutamate oxidase (GLOx) and glutamate dehydrogenase (Gldh). Bacterial cell surface display of Gldh utilising ice nucleation proteins as an anchoring motif has been achieved which showed superior activity, selectivity and thermostability. It was found that the reaction is NADP+ dependent and that the bacteria might be able to generate NADPH. Hence, the Gldh-bacteria has been used as a biosensor for L-glutamate and enzymatically generated NADPH is used for the measurement of signal. Further modification of the biosensor has been done in which the Gldh-bacteria along with multiwalled carbon nanotube is fabricated onto glass carbon electrode. This forms an amperometric sensor which detects L-glutamate at 0.52V (Liang et al. 2015).

Nanotechnology has been combined with cell surface display of enzymes for generation of selective and sensitive biosensor. Xylose dehydrogenase has been displayed on bacterial surfaces along with multi walled carbon nanotubes. The enzyme catalyzes the conversion of xylose to xylonolactone. The XDH-bacteria were then immobilized on an electrode surface in order to develop an electrochemical biosensor which gives out a signal on its activity (Li et al. 2013).

Cell surface display of enzymes that play a role in environment monitoring have also been carried out. On *E. coli* cell surface, an enzyme organophosphorus hydrolase (OPH) has been co-displayed with methyl parathion hydrolase- green fluorescent protein (MPH-GFP). The fluorescence of the GFP that is displayed, is sensitive to changes in pH in the environment and hence can be detected. Hydrolysis of organophosphorus pesticides by the enzyme generates protons which cause a change in pH and get detected by fluorescence (Figure 3). Thus, it is an indication of the organophosphorus pesticide concentration (Liu et al. 2013).

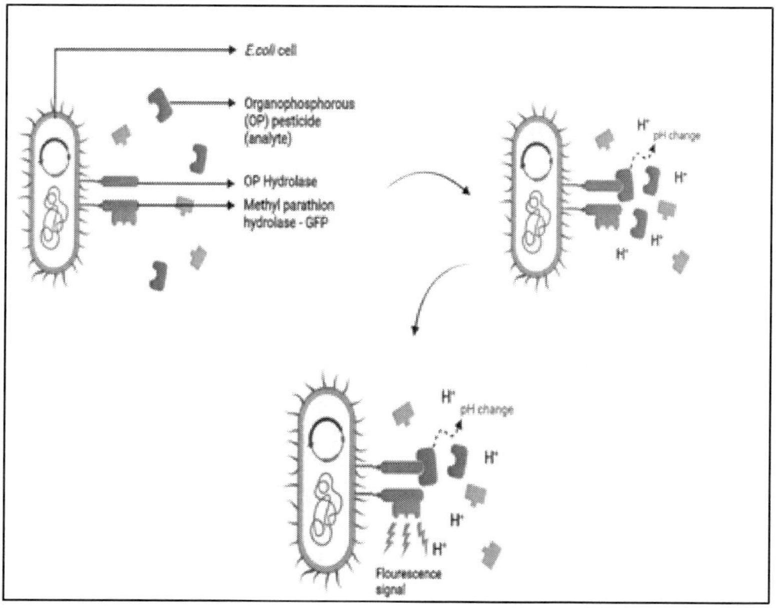

Figure 3. Bacterial cell surface display-based biosensors.

The display of cytokinin oxidase on *E. coli* by making use of the ice nucleation protein from *Pseudomonas borealis* DL7 along with a maltodextrin binding protein as an enhancer has been successful. The enzyme displayed complex is then immobilised onto an electrode to construct a biosensor for selectively detecting cytokinin. This biosensor is used for the detection of hormone levels in plants (Li et al. 2019). Similarly, the oxidoreductase P450 has also been displayed on the surface of cells to develop biosensors and to screen for ligands.

Bacterial laccases have been successfully displayed on the cells of *E. coli* followed by its adsorption on glass carbon electrodes for detecting catechol. The biosensor gives an electrochemical response on its interaction with catechol. This is used in food industry to detect the adulteration of wine and tea samples with catechol (Zhang et al. 2018).

The use of biosensors has also been established to check the level of monosaccharides in food and biomedical analysis. Co-detection of more than one monosaccharide has been achieved by constructing a biosensor having two enzymes displayed simultaneously on a bacterial cell surface. Glucose oxidase and Xylose dehydrogenase have been displayed on a single bacterium and then co-immobilised on multi-walled carbon nanotubes modified electrodes (Li et al. 2013).

Biosensors having spectrometric detection have also been created. An enzyme acetaldehyde dehydrogenase which acts on acetaldehyde, a human carcinogen, has been displayed on a bacterial cell surface and made to react with acetaldehyde present in test samples to produce NADH which gets detected spectrometrically (Liang et al. 2021).

Gold binding peptides displayed on the surface of *E. coli* cells enable the immobilzation of the cells on a gold platform. For this, the anchoring motif FadL has been used. On the same bacteria, to develop a dual display system, another protein – streptavidin is made to display by the anchoring motif of OprF so that it can interact with its target biotin. On flowing the biosensor with HRP, the biotin HRP interaction also gets detected. The detection is carried out by SPR sensing (Park et al. 2009).

A dual reporter-based sensor system has been developed for detecting the presence of heavy metals like zinc and copper in the environment. A metal binding peptide is displayed on *E. coli* outer membrane by anchoring to OmpC protein of the bacteria. Two reporter genes – GFP and RFP are used in the sensor which can fluoresce in the presence of copper and zinc signaling molecules. Thus, real-time multicolour monitoring of heavy metals in the exogenous media is achieved (Ravikumar et al. 2012).

Z domains that bind to IgG molecules have been fused to the outermost membrane of *E. coli* and immobilised on the SPR biosensor. The biosensor detected C-reactive protein levels, an inflammatory biomarker in cardiac disorders (Jeon et al. 2018). Display of Organophosphorus hydrolase (OP) and methyl parathion hydrolase green fluorescence protein (MPH-GFP) on *E. coli* through ice nucleation protein. Specific binding of organophosphorus pesticide molecule with the enzyme OP leads to hydrolysis and production of H+ ions causing a change in pH which then causes a change in the fluorescence intensity which is then detected by spectrophotometer.

4.3. Biosensors Based on Yeast Cell Surface Display

Yeast cell surface display is a eukaryotic expression system which displays functional proteins on the cell surface that are fused to an anchor protein. Table

3 lists some of the most effective yeast cell surface-based biosensors. *Saccharomyces cerevisiae* is the routinely used yeast for display. The selection of an anchor protein is such that it has a signal sequence for promoting transport through the cell onto the cell surface and does not hinder the activity and stability of the protein which is to be displayed. The advantage of the yeast cell surface display over the prokaryotic expression systems is that it allows post translational modifications of the protein. Yeast cell offers a lot of options for cell surface anchor proteins such as Aga1p, Aga2p, Flo1p, etc. All of these proteins belong to the glycosyl-phosphatidylinositol (GPI) family of cell wall proteins (Shusta et al. 2008). The proteins get directed towards the plasma membrane via GPI anchors and then directly link to the cell wall via a β-1,6 glucan bridge for getting incorporated into the cell wall's mannoprotein layer. The protein of interest is fused either to the C-terminal or the N-terminal of the anchor protein which then results in the display of 100000 copies of the fused protein on the cell surface of *S. cerevisiae* (Cherf et al. 2015). The region which is anchored to the cell surface usually is the non-functional region of the protein having desirable activity. One of the most used anchor proteins is Aga2p which has α- agglutinin to whose C-terminus, the protein of interest is attached. Epitope tags are also incorporated in between the fusion regions of the two proteins. The tags such as hemagglutinin tag are incorporated in between the N- terminus of the target protein and Aga2p whereas, the c-myc tag is present at the C-terminus. In *S. cerevisiae*, it is the general secretory pathway that ensures the localisation of the GPI anchored protein on the outermost surface of the cell wall.

Antibodies are used in a large range of biosensor platforms due to the ability of binding with a diverse array of antigens having higher specificity and affinity. Specific traits of a given antibody can be deliberately augmented via yeast cell surface display for enhanced biosensor performance (Siegel RW, 2009). The cell wall protein Aga2p of the yeast has been utilised to be fused with recombinant antibodies and peptide library construction. To monitor the cell surface display and antibody expression on the cell surface, FACS is used.

A glucose monitoring system displaying a fluorescent protein has been developed by Tanaka et al. by making use of yeast cell surface display (Figure 4). The GFP protein was fused to an α-agglutinin protein on *S. cerevisiae* cell surface. The expression of GFP was controlled by a glucose inducible promotor and hence glucose monitoring was made possible. The detection of levels of glucose was carried out by fluorescent measurement of the whole yeast cells. In addition to this, a blue fluorescent protein (BFP) was also surface displayed such that, the decreasing levels of glucose would make the cells fluoresce blue. Thus, in the presence of glucose, the cells showed GFP while in the absence they gave a blue colour (Shibasaki et al. 2001).

Table 3. Biosensors based on yeast cell surface display

S. No	Organism/Peptide	Biosensor	Application	Reference
1	GFP/BFP on *S. cerevisiae*	Fluorescence	Glucose detection	(Shibasaki et al. 2001; Ye et al. 2000)
2	ECFP/EYFP on *S. cerevisiae*	Fluorescence	Ammonium and phosphate ion detection	(Shibasaki et al. 2001)
3	EGFP on *S. cerevisiae*	Fluorescence	Protein expression	(Shibasaki et al. 2003)
4	Glucose oxidase on *S. cerevisiae*	Voltammetry	Glucose detection	(Wang et al. 2013)
5	Anti OmpD	Electrochemical ELISA	Detection of pathogen-Salmonella	(Venkatesh et al. 2015)
6	HCV core protein on *S. cerevisiae*	Optical/ Electrochemical	Detection of anti-HCV core antibodies	(Aronoff-Spencer et al. 2016)
7	AChE on *S. cerevisiae*	Spectrometry, Fluorescence	Pesticide detection	(Liang et al. 2019; Liang et al. 2020)

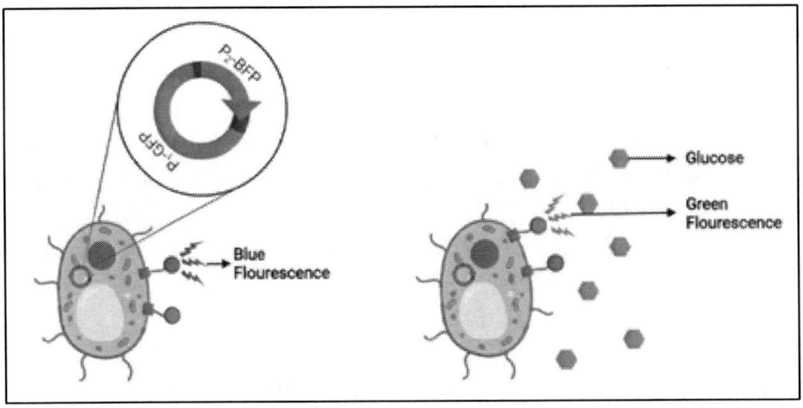

Figure 4. Yeast cell surface display-based biosensors: Glucose biosensor: Yeast cell having a plasmid vector with two genes – GFP and BFP under two different regulatory promotors. The P1-GFP promotor is induced by glucose molecules whereas the P2-BFP promotor is activated under decreased glucose condition. The displayed protein GFP and BFP are bound through α agglutinin anchor motif. In absence of glucose the cells emit blue fluorescence. On binding of glucose with the receptor peptide molecule, the P1-GFP promoter is activated and green fluorescence is emitted.

A similar system has been developed to monitor ammonium and phosphate ions where the expression of the enhanced yellow fluorescent protein and enhanced cyan-blue fluorescent protein was controlled by inducible promotors MEP2 and PHO5 respectively (Ye et al. 2000).

To monitor gene expression, biosensors making use of yeast cell surface display have been used. In these sensors, the enhanced green fluorescence protein (EGFP) is fused with agglutinin on the yeast cell surface and a galactose inducible promotor is used to regulate gene expression. The proteins to be studied were expressed under the same promotor that expressed EGFP. Thus, the biosensor developed enabled real time gene expression studies (Shibasaki et al. 2003).

Glucose oxidase has been surface displayed on yeast cells in order to develop glucose biosensors. The enzyme is fused with the agglutinin anchor protein on *S. cerevisiae* and the cells are immobilised on a glassy carbon electrode for electrochemical measurement by cyclic voltammetry (Wang et al. 2013).

Another biosensor has been developed in which the OmpD antigen of *Salmonella* has been detected by yeast cell-based biosensors having anti OmpD antigen displayed on the cell surface. The anti OmpD protein is displayed using the agglutinin protein as anchor. The interaction is monitored by electrochemical ELISA (Venkatesh et al. 2015).

Yeast cells have been engineered in order to display Hepatitis C Virus (HCV) core proteins along with gold binding peptides on their surface which were able to detect anti HCV antibodies and the detection was monitored either optically or electrochemically (Aronoff-Spencer et al. 2016).

Active acetylcholinesterase has been displayed through the anchor motif agglutinin on the yeast cell surface. The enzyme hydrolyses acetylthiocholine chloride to form thiocholine which in turn reacts with DTNB to absorb light at 415 nm. This biosensor was used to determine the presence of pesticides like paraoxon and parathion, which inhibit the enzyme activity leading to no colorimetric estimation (Liang et al. 2019).

In another study, a fluorescence biosensor was developed using AChE. Due to the action of the enzyme, thiol groups were generated which cause the gold nanoclusters to aggregate via electrostatic interactions. The aggregated nanoparticles led to the quenching of the fluorescence signal (Liang and Han, 2020).

4.4. Biosensors Based on Mammalian Cell Display

The lack of post translational modifications on prokaryotic displayed systems is a major drawback and thus researchers moved on to eukaryotic display systems. The endogenous eukaryotic machinery present in the mammalian cells enables correct folding, thus maintaining the biophysical properties of the protein of interest. However, owing to difficulties such as low transfection efficiency in mammalian cells, the size of the library obtained is less as compared to prokaryotic display libraries (Robertson et al. 2020). In order to study the interaction of human protein-protein, mammalian cell display systems are used (Ivanusic et al. 2020). Endoplasmic reticulum-specific post translational modification and correct folding is required for appropriate biological activity. In some cases, such as display of T cell receptor, single chain variable fragment of an antibody, etc. the need for appropriate environmental conditions is necessary which depends on cell-specific expression and localisation of the protein. Thus, peptide libraries on mammalian cells have been generated for the same. In mammalian cells, the transmembrane receptors are used for linking the peptide of interest which is to be displayed. The N terminus of the CCR5 chemokine receptor, PDGFR or ion channels is used for this purpose (Ivanusic et al. 2020). Higher specificity antibodies have been selected against the Hepatitis B virus from individual B cells immunised with the virus. Such a selection of antibodies is done using FACS. Potential variants of Env, an envelope protein of HIV, displayed on mammalian cells have been screened against for antibodies in order to develop vaccine candidates (Bruun et al. 2017).

The cell membrane components such as ion channel receptors, GPCR, cell surface proteins like heat shock proteins are immobilized on sensor surfaces for detection assays.

A biosensor has been developed in which B cells have been engineered to display antibodies specific to bacteria along with a bioluminescent cytoplasmic protein which is designed such that on the interaction of the receptor-target molecule, signal transduction occurs such that the cytoplasmic protein begins to emit light. The detection of the signal is done by a luminometer (Rider et al. 2003).

Bioelectric recognition assays have been carried out for the inexpensive detection of viral particles in each sample. On the interaction of the virus with the surface receptors, there is a change in the membrane potential of the cell. This change in membrane potential can be detected by fluorescence microscopy (Kintzios et al. 2004).

A mammalian cell-based biosensor has been developed using a membrane formed of polyester as a 3D scaffold for the physical attachment of cells. The cells have been made to express Toll-like receptor 1 proteins on the surface which could interact with pathogenic particles. A secondary antibody having a labeled enzyme induces colorimetric changes which are then detected (Cho et al. 2021).

5. Biosensors Based on *In Vitro* Display Technologies/ Cell Free Biosensors

Recent years have seen huge advances in developing better, cost effective cell-free systems, and having cellular synthetic machineries. The complexity brought forth by the need to have purified proteins that are demanded by biosensors based on antibodies and enzymes has paved the way for researchers to use live biological systems with a self-sustaining capability to synthesize biomolecules capable of biorecognition and use the whole system as a biosensor transducer (Lee and Kim, 2019). It has also been claimed that microbial biosensors offer the opportunity of detecting a wider variety of analytes or targets due to the modularity of their gene expression. In these types of biosensors, the reporter gene and the in-built translation machinery act like a consolidated universal transducer. According to new types of targets, or analytes, a change in gene expression is brought forth in an upstream sequence of DNA, which then bring about different readable signals used in the target recognition in the transducer. These varieties of different sequence elements are constantly made available due to huge advances in the spheres of genetic engineering and synthetic biology. Upon fusing with different reporter genes, biosensors capable of recognising a wide variety of analytes can be developed.

A disadvantage of microbial biosensors that use living microbial cells is that they sometimes have limited and restricted applicability due to cell viability, plasmid incompatibility, analytes' inherent toxicity and the limitation caused by restricted permeability through the cell wall and cell membrane of the analyte into the microbial cell (Iyer and Doktycz, 2014; Pellinen et al. 2004). It has also been reported that whole cell microbial biosensors demand maintenance of the particular cell line and that required frequent passage and cell-culture steps which restrains the applicability of the biosensor.

Thus, a novel biosensor class is now used and is being developed which uses cell-free systems, cell-free protein synthesis (CFPS) which have already been used to produce recombinant, heterologous proteins. In 2013 Catherine et al. reviewed cell-free systems for screening enzymes (Catherine et al. 2013). One commendable advantage of the cell-free system is the greater flexibility offered by the *in vitro* transcription and translation machinery. Due to the *in vitro* nature of the system, the physical, chemical parameters and microenvironment can be modulated and controlled which also helps the researcher to study the functionalities of the macromolecules involved in protein synthesis, like ribosomes and other factors too. They also offer the advantage of being conducted in various phases, like solution phase, within artificial vesicles and also solid phase using gel matrices (Catherine et al. 2013; Griffiths et al. 2006; Noireaux and Libchaber, 2004; Park et al. 2009). Figure 5 represents a cell-free based biosensor, wherein, upon the addition of an analyte to the system, a particular signal is generated which is read and quantified.

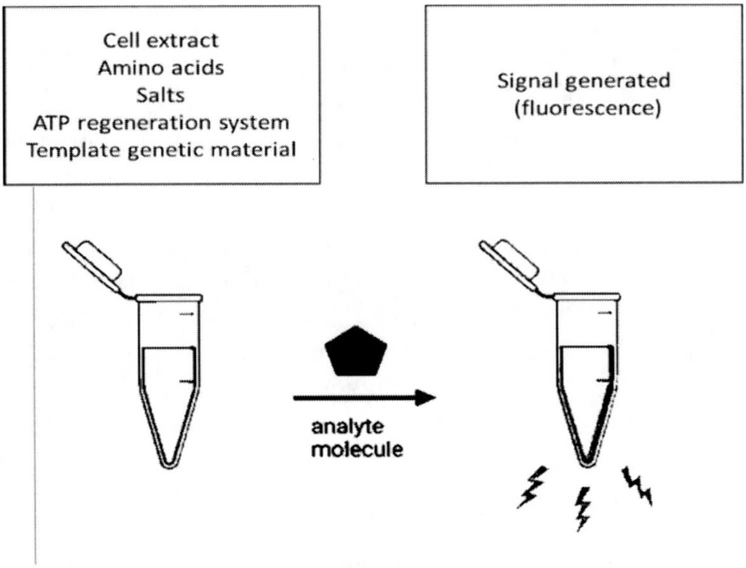

Figure 5. A cell-free based biosensor, which upon addition of the target analyte, confirms its presence with the help of a reporter gene (usually by emitting fluorescence) (Lee and Kim, 2019).

Dopp et al. 2019 reviewed the history and advancements in the research on cell-free protein synthesis systems and the list of ingredients required to be added for proper functioning (Dopp et al. 2019). The three basic ingredients are cell extracts, genetic elements, additives and macromolecules needed to support and nurture the *in vitro* transcription and translation machinery of the *E. coli* based extract. The several unit operations that are involved are given below:

1. Metabolism runs simultaneously and in parallel to keep a constant supply of the energy rich reactants and metabolites.
2. The added genetic element (DNA) is used by the additives to transcribe into several mRNAs
3. Using the translation machinery, the mRNA is then translated to the desired recombinant protein
4. Post-translational modifications are then performed on the protein, like disulfide bonds, or attaching carbohydrate moieties, that are aided by the additives and the cell extract components.

The most used cell extracts are crude cell extracts (S30 fraction) supplemented with mRNA and nucleosides, *E. coli* cell extract (ECE), rabbit reticulocytes lysate (RRL), and wheat germ extract (WGE). The reason they are the most popular choice is due to the presence of high performance translation factors, plus very low proteolytic and nucleolytic activities (Spirin AS, 2004). The crux of continuous cell-free translation systems lies in the continuous inexhaustive supply of reactants and metabolites that need to be consumed by the biological reaction and the removal of the products produced by those same reactions. Many groups have worked on cell-free protein synthesis systems to make them more usable and operable according to ever expanding varieties and numbers of targets and analytes (Spirin et al. 2004; Cai et al. 2015). Cai et al. 2015 noted the improvement of ATP regeneration systems that have been made over the years in cell-free systems. These gradual improvements have greatly increased the productivity of these systems. The first systems used very high energy containing phosphate compounds like Phosphoenolpyruvate to generate ATP, but later on central metabolic pathways like glycolysis and oxidative phosphorylation were used for the same. The advantage of this switch was that now inexpensive compounds like glucose, glutamate, pyruvate could be used for ATP regeneration in the cell free systems.

Soltani et al. 2018 tabulated and reviewed cell-free biosensors specifically and distinctly based on the following three stages of protein synthesis, transcription; translation; and post-translational modifications culminating in protein folding (Soltani et al. 2018).

6. CFPS Biosensing at Transcription

This is designed to inhibit the process of transcribing an indicator protein's DNA sequence unless a target analyte is detected in the system. Repression is brought about when a repressor protein binds to the regulator region of the DNA upstream to the promoter of the indicator gene. However, when the target is present, it binds to the repressor and prevents its binding to the DNA. Upon transcription of the indicator gene, the indicator protein is formed. The common indicator protein used is GFP. Enzymes, due to their catalytic nature, specificity and selectivity for their substrates are also excellent choices as indicator molecules. Reported to yield stronger signals than even fluorescent proteins, they are also the preferred choice since their activity can be tuned finely varying the substrate concentration, physicochemical parameters at the microenvironment level. This fine tunability is greatly helpful if the accuracy and the sensitivity of the biosensor is to be varied. They have been used for several applications, especially in the field of disease diagnostics.

Chappel et al. 2013 developed a transcription based CFPS biosensor employing GFP as the indicator molecule, to detect a quorum sensing molecule released by *Pseudomonas aeruginosa* which frequently infects cystic fibrosis patients (Chappell et al. 2013). It was even reported later that the cell-free transcription-based biosensor had a lower detection limit than the equivalent cell-based biosensor (Wen et al. 2017).

In 2009, another quorum sensing molecule that is used in crosstalk between various bacterial species was verified (Yang et al. 2009).

According to Pellinen et al. 2004, cell-free biosensors employing transcriptional regulation detecting tetracycline hydrochloride (Tn-HCl) and mercuric chloride have a significantly broader detection range and quicker response time than cell-based biosensors (Pellinen et al. 2004). A cell-free biosensor to detect mercury ions was developed using MerR (transcription factor) as its recognition element. Kawaguchi et al. 2020 developed another cell-free biosensor to detect of N-acyl homoserine lactone, a Quorum sensing molecule derived from *Agrobacterium tumifaciens* (Kawaguchi et al. 2008).

7. Nucleic Acid Hybridization-Based Biosensor

To diagnose infectious diseases in patients, the hybridization of the disease-causing organism's genomic material is an important tool for biosensors. For non-infectious diseases, the same phenomenon is used to detect a particular gene or a mutated sequence. It can also be used to assess the abnormal expression of a gene that is occurring in a disease. PCR based detection of nucleic acids is the most popular choice because of its ability to exponentially amplify the template DNA and also offer the choice to detect it in real time, even if the initial concentration of the template is very low. However, due to it requiring skilled personnel and excellent laboratory facilities to handle a PCR based detection assay, researchers are now beginning to diverge into more portable, and accessible and less expensive methods of rapid detection. This technological advancement is welcome in terms of public health and environmental monitoring (Lee and Kim, 2019).

In addition to protein synthesis, RNA molecules, like miRNA and tRNA have other biological functions as well. After the development and diversification of SELEX (systematic evolution of ligands by exponential enrichment), it is now easier to screen synthetic RNAs with a desired function from a library (Ellington and Szostak, 1990).

Paige et al. 2011 developed a method of fluorophore tagging an RNA molecule. They developed a synthetic RNA aptamer molecule, Spinach, that binds to HBI (4-hydroxybenzlidene imidazolinone) (Paige et al. 2011). This compound can mimic the tripeptide (S65-Y66-G67) from GFP, which upon spontaneous cyclization forms HBI. When Spinach binds to HBI, it emits fluorescence, thus enabling visualization of the RNA molecule. Spinach-fluorophore pair is excellent as it has low fluorescence to background noise ratio as it is non-fluorescent before activation and is resistant to photo-bleaching. Spinach based biosensors have been recently used to detect several small molecules like adenosine, ATP, guanine, and even proteins like streptavidin, thrombin. Ying et al. 2018 developed a Spinach based cell free biosensor system where using the presence of specific micro RNAs, the *in vitro* transcription of the spinach gene was activated (Ying et al. 2018). They designed two separate single stranded DNA molecules, having the T7 promoter sequence in one and the Spinach sequence in another, thus prohibiting RNA polymerase to transcribe the gene. However, a miRNA which is complementary to both the abovementioned sequences will bind the two separate ssDNA molecules together by providing a structural base, which can be ligated by ligase, which will bring about transcription by T7 RNA polymerase in the *in vitro* reaction itself.

Dolgosheina et al. 2014 developed another RNA aptamer, called RNA mango which shows high-affinity binding to thiazole orange derivatives (Dolgosheina et al. 2014).

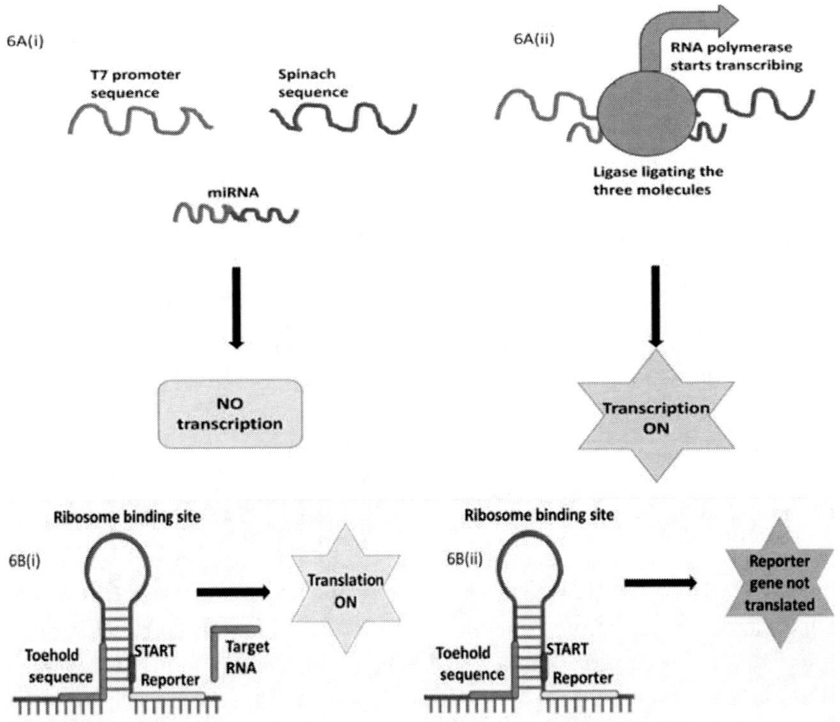

Figure 6. (Continued).

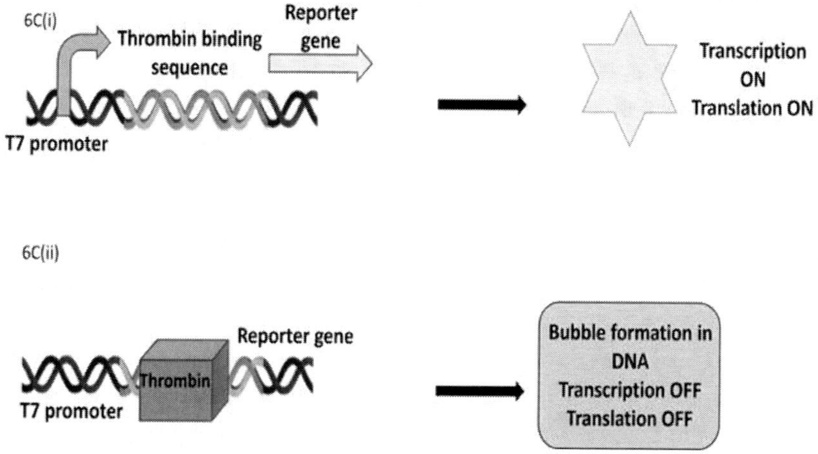

Figure 6. 6A: DNA Hybridization based cell-free biosensors (Pardee et al. 2016): (i) without target RNA, (ii) with target RNA, the toehold seuqnec gets hybridized, opening up the stem loop, thus starting translation. . 6B and C: 6B: Aptamer based cell-free biosensors (Iyer and Doktycz. 2013): (i and ii) without thrombin, the reporter gene gets transcribed and translated, leading to fluorescence, while in 6C(i and ii)in the presence of thrombin, RNA polymerase fails to attach to the promoter, thus inhibiting transcription and consequently translation.

Several biosensors have been developed based on this idea of using transcripts as signaling molecules, but even then, the amplification of signals that is an inherent property of the central dogma is lost. So, there are new avenues followed by researchers to combine the element of translation along with transcription to amplify the biological signal (Lee and Kim, 2019).

Pardee et al. 2016 were credited with developing a synthetic RNA for detecting Zika virus RNA with sensitivity (Pardee et al. 2016). The coding region of the reporter's gene is placed downstream to a toehold sequence in such a way that a hairpin loop is formed by the ribosome binding sequence and the START codon of the reporter gene, thus preventing access to the ribosome and subsequently inhibiting translation. A pre-defined RNA sequence from the Zika virus, against whom the toehold sequence was designed can now be used to hybridize with the latter, thus opening up the hairpin loop, and eventually allowing the reporter gene to be translated (Figure 6).

8. Aptamer Based Cell-Free Biosensor Systems

The need for this arrived when certain analytes could not be detected by any known transcription-based cell-free biosensors. Using SELEX, an aptamer can be obtained with the desired sequence from a combinatorial library, so there is no limit to any target nucleic acid detection by hybridization. If the aptamer-ligand binding activity is devised to be linked with the regulation of a GFP reporter gene expression, a very efficient biosensor can be designed. Iyer and Doktycz, 2013 designed a process wherein they used aptamers to control gene regulation at the transcription level (55). They employed a ssDNA sequence (thrombin binding DNA, or TBA, with a known high affinity towards human α-thrombin) that is recognized and bound by thrombin to insert in their aptamer sequence and inserted it downstream to T7 promoter. If thrombin is present in the system, it gets bound to the aptamer, thus creating a steric hindrance for the RNA polymerase to bind to the promoter and this inhibits transcription. This was reported to have higher sensitivity as compared to the other traditional thrombin detection methods (Lee and Kim, 2019). This aptamer-based biosensor is depicted in figures 6B and 6C.

9. Translational Machinery as Biosensors

The amino acid analysis is indispensable in fields of health, food quality assessment, pharmaceuticals and diagnostics. Jang et al. 2017 developed a cell-free biosensor based on aminoacyl tRNA synthetase (AARS) which is charged with specific amino acids designated for themselves and bind to ribosomes during the process of translation (Jang et al. 2017). The group had developed a rapid and inexpensive method of quantifying amino acids in a mixture, by their polymerization to form a reporter protein, which gives out fluorescence. Unlike the complicacies of an *in vivo* translation event, protein synthesis *in vitro* is done in a homogenous reaction mixture where the contents of the reaction can be individually monitored. By theorizing an incomplete, incompetent cell-free protein synthesis reaction mixture to be analogous to an electrical circuit, it was envisioned that one or few components can be removed from the mixture and different reporter protein outputs can be generated via complementation when we add the missing component. This complementary translational assay was used to measure the concentration of amino acids by Jang et al. 2017. A fluorescent protein, superfolder green fluorescent protein (sfGFP) was synthesized *in vitro* with amino acids, without the ones that need to be quantified. As some of the essential amino acids were not there, the functional protein was not formed, rather a truncated protein was synthesized, so a significant fluorescence signal was not observed. However, if premixed with amino acids that were lacking, fluorescence generated due to sfGFP was observed, which was linearly correlated to amino acids concentration. The validation of the amino acid concentration was carried out by analysing the same in Foetal bovine serum. *E. coli* strain BL21-Star (DE3) cells were used to prepare the cell-free lysate. They reported that this method can be used to determine the concentration of amino acids on site, very rapidly, without any need for chemical treatment or chromatographic steps (Lee and Kim, 2019).

Conclusion

Biosensors constructed based on both cell-based technology and cell free display technologies find numerous applications. The cell-based display approaches are limited by the availability of the anchor protein, its size and also the number of copies displayed on the cell surface. They usually require multiple cloning steps for the synthesis, display and construction of biosensor. On the other hand, the cell free approaches are simple, sensitive and are utilised for the detection of a wide array of analytes. However, the development of *in vitro* cell free display-based biosensors is still in its infancy. With the development of novel molecular biology tools such as better promoters, PCR, sensitive and stable variants of fluorescent proteins, stable and protease free cell lysates, the field is likely to grow further resulting in cost-effective and user-friendly biosensors.

References

Adhikari M, Dhamane S, Hagström AE, Garvey G, Chen WH, Kourentzi K, Strych U, Willson RC. Functionalized viral nanoparticles as ultrasensitive reporters in lateral-flow assays. *Analyst*. 2013 Oct 7;138(19):5584-7.

Aronoff-Spencer E, Venkatesh A, Sun A, Brickner H, Looney D, Hall DA. Detection of Hepatitis C core antibody by dual-affinity yeast chimera and smartphone-based electrochemical sensing. *Biosensors and Bioelectronics*. 2016;86:690-6.

Bratkovič T. Progress in phage display: evolution of the technique and its applications. *Cellular and molecular life sciences*. 2010;67(5):749-67.

Bruun TH, Grassmann V, Zimmer B, Asbach B, Peterhoff D, Kliche A, Wagner R. Mammalian cell surface display for monoclonal antibody-based FACS selection of viral envelope proteins. *MAbs*. 2017 Oct;9(7):1052-1064.

Bunde RL, Jarvi EJ, Rosentreter JJ. Piezoelectric quartz crystal biosensors. *Talanta*. 1998;46(6):1223-36.

Cai Q, Hanson JA, Steiner AR, Tran C, Masikat MR, Chen R, Zawada JF, Sato AK, Hallam TJ, Yin G. A simplified and robust protocol for immunoglobulin expression *in Escherichia coli* cell-free protein synthesis systems. *Biotechnol Prog*. 2015 May-Jun;31(3):823-31.

Catherine C, Lee KH, Oh SJ, Kim DM. Cell-free platforms for flexible expression and screening of enzymes. *Biotechnology advances*. 2013;31(6):797-803.

Chappell J, Jensen K, Freemont PS. Validation of an entirely in vitro approach for rapid prototyping of DNA regulatory elements for synthetic biology. *Nucleic acids research*. 2013;41(5):3471-81.

Charbit A, Boulain JC, Ryter A, Hofnung M. Probing the topology of a bacterial membrane protein by genetic insertion of a foreign epitope; expression at the cell surface. *The EMBO journal*. 1986;5(11):3029-37.

Cherf GM, Cochran JR. Applications of yeast surface display for protein engineering. *Yeast surface display*. 2015:155-75.

Cho IH, Jeon JW, Choi MJ, Cho HM, Lee JS, Kim DH. A Membrane Filter-Assisted Mammalian Cell-Based Biosensor Enabling 3D Culture and Pathogen Detection. *Sensors*. 2021;21(9):3042.

Dolgosheina EV, Jeng SC, Panchapakesan SS, Cojocaru R, Chen PS, Wilson PD, Hawkins N, Wiggins PA, Unrau PJ. RNA mango aptamer-fluorophore: a bright, high-affinity complex for RNA labeling and tracking. *ACS Chem Biol*. 2014 Oct 17;9(10):2412-20.

Dopp JL, Tamiev DD, Reuel NF. Cell-free supplement mixtures: Elucidating the history and biochemical utility of additives used to support *in vitro* protein synthesis in E. coli extract. *Biotechnology advances*. 2019;37(1):246-58.

Ellington AD, Szostak JW. In vitro selection of RNA molecules that bind specific ligands. *Nature*. 1990;346(6287):818-22.

Francisco JA, Campbell R, Iverson BL, Georgiou G. Production and fluorescence activated cell sorting of Escherichia coli expressing a functional antibody fragment on the external surface. *Proceedings of the National Academy of Sciences.* 1993;90(22):10444-8.

Freudl R, MacIntyre S, Degen M, Henning U. Cell surface exposure of the outer membrane protein OmpA of Escherichia coli K-12. *Journal of molecular biology.* 1986;188(3):491-4.

Griffiths AD, Tawfik DS. Miniaturising the laboratory in emulsion droplets. *Trends in biotechnology.* 2006;24(9):395-402.

Hilderbrand SA, Kelly KA, Niedre M, Weissleder R. Near infrared fluorescence-based bacteriophage particles for ratiometric pH imaging. *Bioconjugate chemistry.* 2008;19(8):1635-9.

Ho KC, Chen CY, Hsu HC, Chen LC, Shiesh SC, Lin XZ. Amperometric detection of morphine at a Prussian blue-modified indium tin oxide electrode. *Biosensors and Bioelectronics.* 2004;20(1):3-8.

Ivanusic D, Madela K, Burghard H, Holland G, Laue M, Bannert N. tANCHOR: a novel mammalian cell surface peptide display system. *BioTechniques.* 2020;70(1):21-8.

Ivnitski D, Abdel-Hamid I, Atanasov P, Wilkins E. Biosensors for detection of pathogenic bacteria. *Biosensors and Bioelectronics.* 1999;14(7):599-624.

Iyer S, Doktycz MJ. Thrombin-mediated transcriptional regulation using DNA aptamers in DNA-based cell-free protein synthesis. *ACS synthetic biology.* 2014;3(6):340-6.

Jang YJ, Lee KH, Yoo TH, Kim DM. Complementary cell free translational assay for quantification of amino acids. *Analytical chemistry.* 2017;89(18):9638-42.

Jeon D, Pyun JC, Jose J, Park M. A Regenerative Immunoaffinity Layer Based on the Outer Membrane of Z-Domains Autodisplaying *E. coli* for Immunoassays and Immunosensors. *Sensors.* 2018;18(11):4030.

Jose J. Autodisplay: efficient bacterial surface display of recombinant proteins. *Applied microbiology and biotechnology.* 2006;69(6):607-14.

Karoonuthaisiri N, Charlermroj R, Morton MJ, Oplatowska-Stachowiak M, Grant IR, Elliott CT. Development of a M13 bacteriophage-based SPR detection using Salmonella as a case study. *Sensors and Actuators B: Chemical.* 2014;190:214-20.

Kavita V. DNA biosensors-a review. *J Bioeng Biomed Sci.* 2017; 7(2):222.

Kawaguchi T, Chen YP, Norman RS, Decho AW. Rapid screening of quorum-sensing signal N-acyl homoserine lactones by an in vitro cell-free assay. *Applied and Environmental Microbiology.* 2008;74(12):3667-

Kintzios S, Bem F, Mangana O, Nomikou K, Markoulatos P, Alexandropoulos N, Fasseas C, Arakelyan V, Petrou AL, Soukouli K, Moschopoulou G, Yialouris C, Simonian A. Study on the mechanism of Bioelectric Recognition Assay: evidence for immobilized cell membrane interactions with viral fragments. *Biosens Bioelectron.* 2004 Nov 1;20(4):907-16.

Koopaee H, Rezaei V, Esmailizadeh A. DNA Biosensors Techniques and Their Applications in Food Safety. Environmental Protection and Biomedical Research: A mini-review. *J Cell Dev Biol.* 2020;3(1):28-35.

Lee HE, Kang YO, Choi SH. Electrochemical-DNA biosensor development based on a modified carbon electrode with gold nanoparticles for influenza A (H1N1) detection:

effect of spacer. *International Journal of Electrochemical Science*. 2014;9(12):6793-808.

Lee KH, Kim DM. In vitro use of cellular synthetic machinery for biosensing applications. *Frontiers in pharmacology*. 2019;10:1166.

Lee SW, Lee KY, Song YW, Choi WK, Chang J, Yi H. Direct electron transfer of enzymes in a biologically assembled conductive nanomesh enzyme platform. *Advanced Materials*. 2016;28(8):1577-84.

Li L, Liang B, Li F, Shi J, Mascini M, Lang Q, Liu A. Co-immobilization of glucose oxidase and xylose dehydrogenase displayed whole cell on multiwalled carbon nanotube nanocomposite films modified electrode for simultaneous voltammetric detection of D-glucose and D-xylose. *Biosens Bioelectron*. 2013 Apr 15;42:156-62.

Li Y, Gong B, Liang X, Wu Y. Direct electrochemistry of bacterial surface displayed cytokinin oxidase and its application in the sensitive electrochemical detection of cytokinins. *Bioelectrochemistry*. 2019;130:107336.

Liang B, Han L. Displaying of acetylcholinesterase mutants on surface of yeast for ultratrace fluorescence detection of organophosphate pesticides with gold nanoclusters.

Liang B, Li L, Tang X, Lang Q, Wang H, Li F, Shi J, Shen W, Palchetti I, Mascini M, Liu A. Microbial surface display of glucose dehydrogenase for amperometric glucose biosensor. *Biosens Bioelectron*. 2013 Jul 15;45:19-24.

Liang B, Liu Y, Zhao Y, Xia T, Chen R, Yang J. Development of bacterial biosensor for sensitive and selective detection of acetaldehyde. *Biosensors and Bioelectronics*. 2021;193:113566.

Liang B, Wang G, Yan L, Ren H, Feng R, Xiong Z, Aihua Liu. Functional cell surface displaying of acetylcholinesterase for spectrophotometric sensing organophosphate pesticide. *Sensors and Actuators B: Chemical*. 2019;279:483-9.

Liang B, Zhang S, Lang Q, Song J, Han L, Liu A. Amperometric L-glutamate biosensor based on bacterial cell-surface displayed glutamate dehydrogenase. Analytica chimica acta. 2015;884:83-9. *Biosensors and Bioelectronics*. 2020;148:111825.

Liu R, Yang C, Xu Y, Xu P, Jiang H, Qiao C. Development of a whole-cell biocatalyst/biosensor by display of multiple heterologous proteins on the *Escherichia coli* cell surface for the detoxification and detection of organophosphates. *Journal of agricultural and food chemistry*. 2013;61(32):7810-6.

Mehrvar M, Bis C, Scharer JM, Moo-Young M, Luong JH. Fiber-optic biosensors-trends and advances. *Analytical sciences*. 2000;16(7):677-92.

Nanduri V, Bhunia AK, Tu SI, Paoli GC, Brewster JD. SPR biosensor for the detection of L. monocytogenes using phage-displayed antibody. *Biosensors and Bioelectronics*. 2007;23(2):248-52.

Noireaux V, Libchaber A. A vesicle bioreactor as a step toward an artificial cell assembly. *Proceedings of the National Academy of Sciences*. 2004;101(51):17669-74.

Paige JS, Wu KY, Jaffrey SR. RNA mimics of green fluorescent protein. *Science*. 2011;333(6042):642-6.

Pancrazio J, Whelan J, Borkholder D, Ma W, Stenger D. Development and application of cell-based biosensors. *Annals of biomedical engineering*. 1999;27(6):697-711.

Pardee K, Green AA, Takahashi MK, Braff D, Lambert G, Lee JW, Ferrante T, Ma D, Donghia N, Fan M, Daringer NM, Bosch I, Dudley DM, O'Connor DH, Gehrke L, Collins JJ. Rapid, Low-Cost Detection of Zika Virus Using Programmable Biomolecular Components. *Cell.* 2016 May 19;165(5):1255-1266.

Park TJ, Zheng S, Kang YJ, Lee SY. Development of a whole-cell biosensor by cell surface display of a gold-binding polypeptide on the gold surface. *FEMS microbiology letters.* 2009;293(1):141-7.

Pellinen T, Huovinen T, Karp M. A cell-free biosensor for the detection of transcriptional inducers using firefly luciferase as a reporter. *Analytical biochemistry.* 2004;330(1):52-7.

Peng H, Chen IA. Rapid colorimetric detection of bacterial species through the capture of gold nanoparticles by chimeric phages. *ACS nano.* 2018;13(2):1244-52.

Pividori M, Merkoci A, Alegret S. Electrochemical genosensor design: immobilisation of oligonucleotides onto transducer surfaces and detection methods. *Biosensors and Bioelectronics.* 2000;15(5-6):291-303.

Rashid JIA, Yusof NA. The strategies of DNA immobilization and hybridization detection mechanism in the construction of electrochemical DNA sensor: A review. *Sensing and bio-sensing research.* 2017;16:19-31.

Ravikumar S, Ganesh I, Yoo IK, Hong SH. Construction of a bacterial biosensor for zinc and copper and its application to the development of multifunctional heavy metal adsorption bacteria. *Process Biochemistry.* 2012;47(5):758-65.

Rider TH, Petrovick MS, Nargi FE, Harper JD, Schwoebel ED, Mathews RH, Blanchard DJ, Bortolin LT, Young AM, Chen J, Hollis MA. A B cell-based sensor for rapid identification of pathogens. *Science.* 2003 Jul 11;301(5630):213-5.

Robertson N, Lopez-Anton N, Gurjar SA, Khalique H, Khalaf Z, Clerkin S, Leydon VR, Parker-Manuel R, Raeside A, Payne T, Jones TD, Seymour L, Cawood R. Development of a novel mammalian display system for selection of antibodies against membrane proteins. *J Biol Chem.* 2020 Dec 25;295(52):18436-18448.

Shibasaki S, Ninomiya Y, Ueda M, Iwahashi M, Katsuragi T, Tani Y, Harashima S, Tanaka A. Intelligent yeast strains with the ability to self-monitor the concentrations of intra- and extracellular phosphate or ammonium ion by emission of fluorescence from the cell surface. *Appl Microbiol Biotechnol.* 2001 Dec;57(5-6):702-7.

Shibasaki S, Tanaka A, Ueda M. Development of combinatorial bioengineering using yeast cell surface display—order-made design of cell and protein for bio-monitoring. *Biosensors and Bioelectronics.* 2003;19(2):123-30.

Shusta EV, Pepper LR, Cho YK, Boder ET. A decade of yeast surface display technology: where are we now? *Combinatorial chemistry & high throughput screening.* 2008;11(2):127-34.

Siegel RW. Antibody affinity optimization using yeast cell surface display. *Biosensors and Biodetection*: Springer; 2009. p. 351-83.

Singh R, Verma R, Sumana G, Srivastava AK, Sood S, Gupta RK, Malhotra BD. Nanobiocomposite platform based on polyaniline-iron oxide-carbon nanotubes for bacterial detection. *Bioelectrochemistry.* 2012 Aug;86:30-7.

Soltani M, Davis BR, Ford H, Nelson JAD, Bundy BC. Reengineering cell-free protein synthesis as a biosensor: Biosensing with transcription, translation, and protein-folding. *Biochemical Engineering Journal*. 2018;138:165-71.

Spirin AS. High-throughput cell-free systems for synthesis of functionally active proteins. *Trends in biotechnology*. 2004;22(10):538-45.

Takakusagi Y, Takakusagi K, Sugawara F, Sakaguchi K. Use of phage display technology for the determination of the targets for small-molecule therapeutics. *Expert opinion on drug discovery*. 2010;5(4):361-89.

Vasuki, S & Varsha, V & Mithra, R & Dharshni, S & Abinaya, R & Dharshini, N & Sivarajasekar, N.. (2019). *Thermal biosensors and their applications*.

Venkatesh AG, Sun A, Brickner H, Looney D, Hall DA, Aronoff-Spencer E. Yeast dualaffinity biobricks: Progress towards renewable whole-cell biosensors. *Biosensors and Bioelectronics*. 2015;70:462-8.

Wang H, Lang Q, Li L, Liang B, Tang X, Kong L, Mascini M, Liu A. Yeast surface displaying glucose oxidase as whole-cell biocatalyst: construction, characterization, and its electrochemical glucose sensing application. *Anal Chem*. 2013 Jun 18;85(12):6107-12.

Wang J, Xu D, Kawde AN, Polsky R. Metal nanoparticle-based electrochemical stripping potentiometric detection of DNA hybridization. *Analytical chemistry*. 2001;73(22):5576-81.

Wang L, Chen X, Wang X, Han X, Liu S, Zhao C. Electrochemical synthesis of gold nanostructure modified electrode and its development in electrochemical DNA biosensor. *Biosensors and Bioelectronics*. 2011;30(1):151-7.

Wen KY, Cameron L, Chappell J, Jensen K, Bell DJ, Kelwick R, Kopniczky M, Davies JC, Filloux A, Freemont PS. A Cell-Free Biosensor for Detecting Quorum Sensing Molecules in *P. aeruginosa*-Infected Respiratory Samples. *ACS Synth Biol*. 2017 Dec 15;6(12):2293-2301.

Yang YH, Kim TW, Park SH, Lee K, Park HY, Song E, Joo HS, Kim YG, Hahn JS, Kim BG. Cell-free *Escherichia coli*-based system to screen for quorum-sensing molecules interacting with quorum receptor proteins of Streptomyces coelicolor. *Appl Environ Microbiol*. 2009 Oct;75(19):6367-72.

Ye K, Shibasaki S, Ueda M, Murai T, Kamasawa N, Osumi M, Shimizu K, Tanaka A. Construction of an engineered yeast with glucose-inducible emission of green fluorescence from the cell surface. *Appl Microbiol Biotechnol*. 2000 Jul;54(1):90-6

Ying ZM, Tu B, Liu L, Tang H, Tang LJ, Jiang JH. Spinach-based fluorescent light-up biosensors for multiplexed and label-free detection of microRNAs. *Chemical Communications*. 2018;54(24):3010-3.

Zhang Y, Dong W, Lv Z, Liu J, Zhang W, Zhou J, Xin F, Ma J, Jiang M. Surface Display of Bacterial Laccase CotA on Escherichia coli Cells and its Application in Industrial Dye Decolorization. *Mol Biotechnol*. 2018 Sep;60(9):681-689.

Chapter 2

Sensing the Environmental Contaminants: Current Trends and Future Aspects of Microbial-Derived Biosensors

Himanshu Bariya
Ashish Patel
and Shreyas Bhatt*
Department of Life sciences, Hemchandracharya North Gujarat University, Patan, Gujarat, India

Abstract

Microorganisms are uniquely suited to design and fabrication as bio-sensing materials because of their superior capacity to sense a variety of environmental stimuli. Microbes are appropriate species because simple and practical techniques are available for DNA modification and culture preservation. In addition, microbial biosensors are compact and user friendly in nature and cost-effective. Microbes' distinctive functional and physiological characteristics make them a more suitable candidate as compared to other spp. Various traits of typical biosensors including vulnerability, selectivity and range of target can be improved through controlling gene alteration. The utilization of genetically modified (GM) microbe-based sensors leads to the production of biosensors with exceptional potential for modern applications. These modified biosensors, enable the development of portable device that can detect a broad range of environmental contaminants with ease.

In this chapter, we reassess the present development scenario related to the role of biosensors along with microbes for real-time sensing of a

* Corresponding Author's Email: sabhatt9@gmail.com.

In: Biosensing
Editor: Rushika Patel
ISBN: 979-8-88697-911-4

different range of contaminants in the environmental niches. The article sheds some light on the potential use of microbial biosensors with altered genetic information for monitoring environmental contaminants and pollutants using state-of-the-art techniques. The chapter also explores the most discussed societal and technical issues to identify appropriate solutions with the help of microbial biosensors, ultimately increasing their acceptability. Finally, the chapter concludes with suggestions for future research, including a focus on microbe-based biosensors that can detect a wide range of contaminants or hazardous substances across a broad spectrum in a synchronized manner.

Keywords: microbial-based biosensors, environmental contaminants, genetically modified microbes, regulatory genes

1. Introduction

Plants Pollution of the environment is one of the most important global issues, resulting from anthropogenic activities and the buildup of diverse environmentally hazardous pollutants discharged by the industry as well as current urbanization and population growth. Environmental pollutants come in a variety of forms and are abundantly present in the surroundings disrupting all living systems, including humans (Xiong et al., 2022). As a result, environmental protection and safety have become a major global concern. Because the control of hazardous chemicals is a key scenario for pollution cleanup, environmental researchers are focused on discovering effective solutions for tracking environmental contaminants. To detect pollutants, traditional analytical methods such as chromatographic (Li et al., 2021; Deng et al., 2021; Wu et al., 2022) and spectroscopic (Brunnbauer et al., 2021; Trapananti et al., 2021; Dhote et al., 2013; Murzyn et al., 2021) approaches are used. These analytical procedures, on the other hand, are time-consuming, need numerous sample processing stages, may use hazardous compounds, and are arduous; and the instruments requires skilled personnel. The necessity for accurate, cost-effective, real-time analytical equipment for quick detection and analysis of harmful components has resulted in the development of new biosensing devices to address the problem (Gavirls et al., 2022).

Biosensors are devices capable of detecting and identifying signals that are emitted by or originate from cells or tissues. They have accepter elements and several types of transducers (Neethirajan et al., 2018). Biosensors are divided into a variety of categories based on the cellular, molecular and tissue components they contain (Hernandez-Vargas et al. 2018). Figure 1 depicts the working principle of the biosensor. The receiving elements of molecular biosensors with great selectivity are biologically active molecules such as proteins, nucleotides and biofilms (Du et al., 2005, Wang et al., 2014).

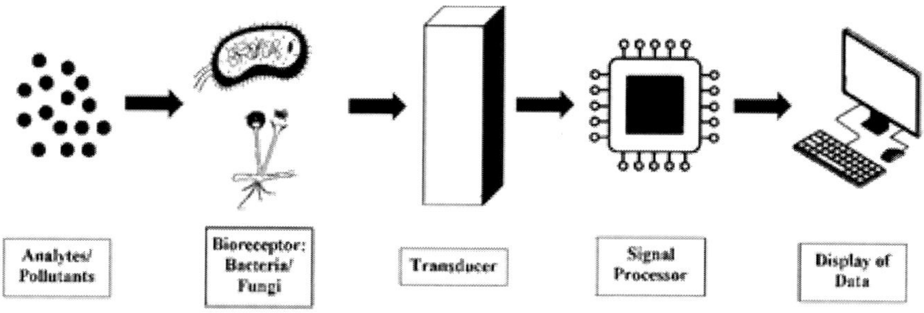

Figure 1. Working principle of Biosensor.

Due to various limitations, such as restricted sensing ability, expensive macromolecule isolation procedures, and a shorter functional lifespan of identifying molecules, molecular biosensors are vulnerable to real-time exploitability (Daunert et al., 2000). Sensors made from whole cells/cells or complete tissue, on the other hand, have made rapid development in terms of novel fabrication processes and immobilization. A breakthrough in high-end technology has given these tissue- or whole-cell-based biosensors special and undiscovered benefits. (Liu et al., 2014). Environmental signals are less sensitive to whole cell biosensors (WCB) or microbial cell biosensors (MCB) than molecular biosensors. However, genetic engineering techniques can be used to create biosensors that can sense an array of signals inside the cell (Xu and Ying, 2011). The choice of the reporter gene, as well as the affinity and screening of molecular recognition, which occurs when regulator proteins bind to their target molecules, are important elements in MCB performance (Raut et al., 2012). The reporter gene used, as well as the affinity and screening of molecular recognition, which happens when regulator proteins join to their target molecules, is all critical factors in MCB performance. The MCB's working route entails the recognition of specific molecules, which are then amplified into a signal via a processor. Encapsulation and the utilization of living cell or bacterial cell to generate recognition molecules are used to measure this readout process (Gui et al., 2017). MCBs are more sensitive to environmental signals and can detect a wide range of chemicals than conventional biosensors due to well-worse genetic modification and working under varied settings, such as diverse temperatures and pH (Tian et al., 2017).

Due to their unique properties such as high selectivity, vulnerability, and high-throughput *in-situ* sensing capability, MCBs have been successfully used in a variety of industries including pharmaceutical, environmental as well as food analysis, and drug testing (Raut et al., 2012). Chemicals are widely used in both the industrial and agricultural sectors, results in higher quantities of hazardous pollutants in the environment (Hernandez-Vargas et al., 2018; Peng et al., 2018; Bilal et al., 2019a). The buildup of such dangerous substances in the environment has negative consequences for both ecosystem and human health (Barrios-Estrada et al., 2018; Bilal et al., 2018a; 2018b). As a result, in order to carry out pollution comfort applications, real-time, consistent, susceptible, and cost-effective surveillance of such harmful contaminants is critical (Rasheed et al., 2018a; 2018b; Bilal et al., 2019b). Traditional methods for displaying environmental toxicants, such as chromatographic procedures, need expensive and specialized units, difficult methodologies, time, and knowledge (Liu et al., 2010). Whole microbial cell- based biosensors tend to be more ideal for the detecting and monitoring of environmental linked hazardous pollutants due to their associated benefits of superior analytical properties such as specificity, sensitivity and limit of detection (LOD) (Maduraiveeran and Jin, 2017). This review examines the current context for designing and developing MCBs, as well as their various applications for effective environmental monitoring and detection. Also, many types of regulatory proteins and reporter genes which play a crucial role in the effective functioning of fabricated biosensors are also mentioned. Global issues and future remarks on biosensors for effective sensing of a variety of environmental pollutants are also focused on.

2. Selection of Specific Microorganism

The selection of microorganisms is the most important criteria for the successful development and fabrication of functional biosensors. The selected microorganisms should contain high affinity with their complementary transducer for sensing signals from surrounding environments. In many cases, yeast and bacteria are usually exploited as biosensing organisms (l. c.). Moreover, the selected microorganism ought to be strong and able to sense distinctive pollutants even in minute quantities, to make sure rate-efficient detection. In last few years whole-cell biosensors and microbial fuel cells have opened new avenues within the discipline of environmental pollutants sensing in which genetic engineering additionally have played an important role

(Mulchandani and Rajesh, 2011; Anu Prathap et al., 2012; Liu et al., 2014; Ayyaru and Dharmaligman, 2013; Nigam and Shukla 2015; Kumar et al., 2020). We will improve or manipulate the genetic attributes of organisms to improve to a greater extent the precise mechanisms of contaminants sensing or explicit them in another organism (Mulchandani and Rajesh, 2011). Model and experimental organisms with specific growth circumstances, such as Saccharomyces cerevisiae and Escherichia coli, may be combined with specific DNA sections or genes coding for detecting traits. The accurate aggregation of biorecognition detail with its transducer assets should be required to accomplish the first-rate promising detection of pollution.

3. Configuration of Biosensor

In general, three predominant kinds of microbe-based biosensors which are primarily based on their precise types of transducers (l. c.) have been developed. Principles of optical measurements which include fluorescence, bioluminescence and colorimetry for the conversion of a biological signal into a precise signal output are extensively used for the functioning of optical biosensors (l. c.). Figure 2 illustrates the class and subclasses of different types of biosensors. Taking an advantage of genetic engineering, incorporation of

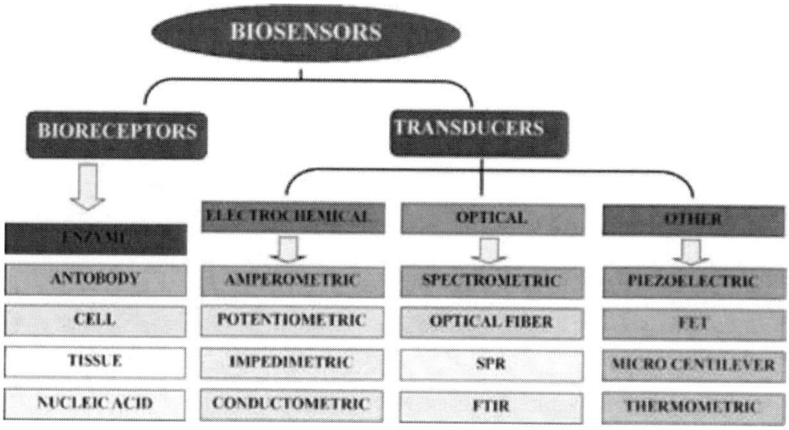

Figure 2. Biosensor classification.

DNA sequences or gene encoding for bioluminescence and fluorescence inside the target host can be achieved. Research report shows the usage of luciferase enzyme (Niazi et al., 2008; Shin, 2010; Chan et al., 2013) and green fluorescent protein (GFP) utilization (Kim et al., 2015) for bioluminescence and fluorescence detection respectively, which can be detected as a specific signal with the aid of biosensor detector.

Differential electric potential, current and conductivity are the operative principles for the electrochemical transducers, resulting from microbes-pollutant interactions. They are furthermore sub-divided into three categories of biosensors. In amperometric microbial biosensors, specific stable potential along with reference electrode are used and an oxidation-reduction reaction on the surface of the electrode produced corresponding electric potential. (l. c.). The working pattern of amperometric microbial biosensors has been nicely elaborated by means of Yong et al. (2011) and Wang et al. (2013); and Mulchandani and Rajesh (2011) demonstrated the principle and working of the potentiometric transducer. Transducers use the potentiometric principle and ion specific electrodes to turn biological signals into electrical signals that can be detected. Although, these kinds of transducers are non-prone, they create large errors and a non-significant relationship between the quantity of pollutants and area of signal peak (l. c.). On the other hand, biosensors working on the conductometric principle detect any differences in conductivity of the selected environmental samples like water, air and soil, as a result of the pollutants contamination in these samples. Microbial Fuel Cells (MFC) are bio-electrochemical gadgets that generate electrical potential by the routing electrons generated by the oxidation of organic material at the anode site to the high energy compounds (oxidized) such as oxygen at the cathode site via an external circuit. (Di Lorenzo et al., 2014). Figure - 3 depicts the working pattern of MFC. The electrons from the anode can be transported to the cathode located on the opposite end with the assistance of an external circuit. In the cathode chamber, water is generated from oxygen and proton after passing through a proton exchange membrane. (Du et al., 2007). The analyte can use this electric current generated by the oxidation reaction as a transducer to create microbial responses. MFC has a simple layout and low cost, which makes it used extensively in various environmental applications. These types of MFCs can easily measure Biological Oxygen Demand (BOD) (Ayyaru and Dharmaligman, 2014) and effectively sense the contamination of heavy metals in the surroundings (Di Lorenzo et al., 2014). Simple MFC are perfect for creating unique, confirmative, and sensitive biosensors since they may contain a variety of microbial populations.

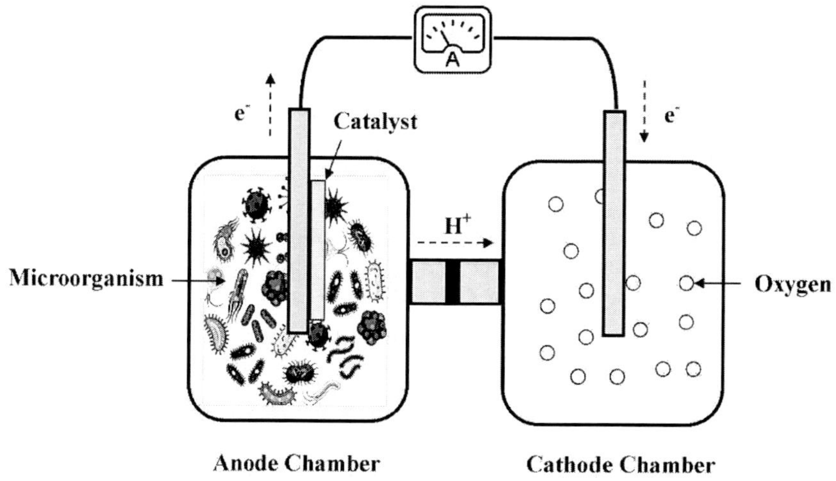

Figure 3. Microbial fuel cell.

4. Microbial Cell Biosensor (MCB)

As biorecognition elements, microbial cell biosensors make use of natural or genetically modified microorganisms such as bacteria, fungi, algae, and yeast, which discharge detectable signals, that can be quantified by an associated transducer. Worldwide research on biosensors application explores several ways on the usage of MCBs for environmental assessment for the presence of pesticide residues, hazardous heavy metals and diverse pollutants/contaminants. Even at extremely low quantities, heavy metals such as cadmium (Cd), mercury (Hg) and lead (Pb) in the environment have severe deleterious effects on living species. Heavy metal biosensing is one of the most significant applications of MCBs. (Wang et al., 2014). A pathogenic bacterium Staphylococcus aureus has been used as a biorecognition element in an MCB to study the effect of a range of cadmium (II) ion concentration on S. aureus growth and metabolism (Sochor et al. 2011). Rajkumar et al. (2017) used MCBs for organophosphate pesticide biosensing, which is the most dangerous environmental pollutant. MCBs have a few advantages, including high sensitivity, selectivity, and detection limits, making them useful in a variety of disciplines. Table 1 shows a list of MCBs used for analytes/contaminants significant to environmental and agricultural monitoring. (Moraskie et al., 2021).

Table 1. Microbial whole-cell biosensor reporters. Adopted from Moraskie et al., 2021

Reporter Protein	Gene	Detection method	Advantages	Disadvantages
Chloramphenicol (CM) acetyltransferase	Cat	FL, RI	No endogenous activity	Radioisotope/fluorescent labeling, requires substrate and cofactor addition, narrow linear range
β -galactosidase	lacZ	FL, CL, EC, CR	Numerous detection methods, high stability at various temperatures (Asraf and Gunasekaran, 2010), detection by naked eye, applicable in anaerobic environment	Exogenous substrate requirement
Bacterial luciferase	lux	BL	Rapid response, high signal to noise ratio, No endogenous activity	Oxygen requirement, thermal lability
Firefly luciferase	Fluc	BL	High signal to noise ratio, No endogenous activity	Exogenous substrate, oxygen and ATP requirement
Renilla luciferase	Rluc	BL	High signal to noise ratio	Exogenous substrate and oxygen requirement
NanoLuc luciferase	Nluc	BL	High intensity, no thermal lability, ATP independent reaction	Exogenous substrate and oxygen requirement
Aequorin	aequorin	BL	High signal to noise ratio, no endogenous activity	Exogenous substrate and Ca2+ requirement
Green fluorescent protein (GFP)	avGFP	FL	No substrate requirement, high stability, numerous mutated versions available such as CFP, BFP, YFP, etc.	Background fluorescence, slow maturation, low sensitivity
Red fluorescent protein (DsRed)	DsRed	FL	No substrate requirement, high stability, low background fluorescence, wide selection of mutated DsRed proteins available such as mCherry, and mBanana	Slow maturation, moderate sensitivity
Uroporphyrinogen III methyltransferase	cobA	FL	No substrate requirement, auto- fluorescent	Endogenous activity
Catechol 2,3dioxyge- nase or Metapyrocatechase (MPC)	xylE	CR	Instrument-free detection, detection by naked eye	Exogenous substrate required
Chromoprotein (amilCP, CjBlue, eforRed, aeBlue)	amilCP, eforRed, aeBlue	CR	Instrument-free detection, detection by naked eye, No substrate requirement	Oligomer formation

FL=Fluorescence, CL=Chemiluminescence, RI = Radioisotope, BL=Bioluminescence, CM=Chemiluminescence, CR=Colorimetric, EC = Electrochemical.

4.1. Regulatory and Reporter Gene for MCBs

The functional potential of WCBs is influenced by both reporter genes and regulatory proteins. Many types of biosensors incorporate with genetic codes have been well studied and reported globally, according to a literature review (Johnson et al., 2017; Xu, 2018). By combining sensitive genes to biorecognition genes, genetically encoded microbes-derived biosensors are crafted. The recombinant genes can then be directly inserted into the chromosome or via a plasmid into a microbial host. The regulator activates the promoter region in the presence of the target drug compound, resulting in reporter gene transcription and a detectable signal. The interactions of regulatory genes to their target analytes determine the specificity and sensitivity of microbial biosensors (Shin, 2011). Regulatory proteins are typically activated by structurally identical molecules that conformationally bind to their effectors' binding active sites. For several years, researchers in the field of biosensors have been trying to figure out how regulatory proteins interact with effectors to activate promoters. The molecular mechanism by which regulatory proteins performed functions, such as signal detection and promoter elicitation, has been discovered by observation of different aromatic pollutants degrading bacterial regulator-promotor pair (Diaz and Prieto, 2000). The first laptop-assisted systemic method was utilised to demonstrate regulator specificity was in 2003. (Looger et al. 2003). This type of research can also aid scientists in the development of engineering regulators, which can affect a wide range of target molecules. The reporter genes are found inside the cell cytoplasm. These reporter genes can translate microbe-derived inputs into a measured output, and the direction of mobility is critical for the MCBs to have increased selectivity and sensitivity. Once the regulatory proteins become activated in the influence of the signal, the microbial biosensors respond positively, followed by production of the reporter gene product. The obtained signal, which includes bioluminescence, fluorescence, and colour change, may be visible to the naked eye (Johnson et al., 2017). Today, a wide range of reporter genes are easily accessible and are being used in the development and production of biosensors (see Table 2) (Moraskie et al, 2021).

Table 2. Whole-cell biosensors for use in environmental and agricultural monitoring. Adopted from Moraskie et al., 2021

Name of WCBs	Analytes/ Pollutants	Promoter/Receptor Construct
Nostoc sp. PCC 7120	N (NH^{+3}or NO^{-3})	*glnA/nir/gifA-luxCDABE*
S. elongatus PCC 7942		*glnA/nir/gifA-luxCDABE*
E. coli		*glnAp2' -luxCDABE*
E. coli	P (PO^{-3})	*phoA`-luxCDABE*
E. coli XL1-Blue	Cu2+	PcusC–rfp
E. coli TOP10		*cusR*-*PcusC-gfp*
C. metallidurans		*copSR*-*PcopT-rfp*
C. metallidurans		*copSR*-*PcopQ-MjDOD*
Wild-type and mutant *E. coli* BL21		PcopA-*EGFP* pCDF-Duet-*CueR*
E. coli XL1-Blue	Zn2+	PzraP–gfp
B. megaterium strain WH320		*smtB-EGFP*
P. putida X4		P*czcR3-EGFP*
E. coli TOP10	As3+	*arsR*-*ParsR-gfp*
E. coli		*arsR-O/P-luxCDABE*
E. coli DH5α		*ars-pr-mCherry*
A. baylyi	Cr^{6+}	P*recA-luxCDABE*
E. coli		*recA-luxCDABE*
D. radiodurans	Cd^{2+}	pRADI-P0659-1-crtI
P. putida 06909		*gfp-lacI^q*-*PcadR-Ptac-cadR*
*Synechocystis*sp. PCC 6803	Ni2+	*nrsR*-*PnrsBACD-luxAB*
E. coli DH5α	Pb2+	*ΔpbrA*-*PpbrRT-pbrR-gfp*
P. putida	Hg2+	*merR-egfp*
*Synechocystis*sp. PCC 6803	Co2+	*coaR*-*PcoaT-luxAB*
B. subtilis ars-23	Sb_2S_3	*arsR-lacZ*
*B. sartisoli*RP007	Naphthalene	*phn-luxAB*
E. coli DH5α	BTEX	*LuxAB-TbuT*
E. coli K12 strain MG1655	2,4-DNT	*yqjF-luxCDABE*
E. coli K12 strain MG1655	2,4,6-TNT	*ybiJ-luxCDABE*
A. tumefaciens	C_{14}-HSL	*traG-lacZ-TraR*
E. coli DH5α	$3OC_6$-HSL	P*nptII-gfp/ahlR/T1—4/PahlImcherry*
E. coli DH10B	3-PBA	p*AmilCP*_J104

4.2. Genetically Engineered Strains Derived MCBs

Several bacteria have been produced using recombinant DNA technology, with native regulatory genes coding for a bio-receptor gene and transcriptional regulator, as well as a promoter, generally assimilated. (Vollmer et al., 2004; Ron, 2007). The most common and widely applicable reporter genes are *lacZ* (producing galactosidase), *lux/luc* (expressing firefly/bacterial luciferase enzyme), and *gfp* (encoding green fluorescent protein). Previous studies have shown that genetically modified (GM) bacteria can be used to detect and monitor environmental pollutants (Hansen and Srensen, 2001; Keane et al., 2002; Belkin, 2003). For the detection and sensing contamination of benzene in air samples taken at the site of oil refinery, Gennaro et al. (2011) used two GM E. coli strain-based systems. These recombinant GM E. coli strains contain two genes from P. putida MST, one gene encodes for Benzene Dioxygenase (BED) and the other gene encode for Benzene Dihydrodiol Dehydrogenase (BDDH). The previous enzyme (BED) catalysed benzene in to dihydrodiol, followed by removal of hydrogen by the BDDH enzyme for the manufacture of catechol. The results showed that the efficiency of newly engineered GM E. coli cells was sufficient to detect benzene fumes at very less concentration of 0.01 mM within a short amount of time. The insertion of a specific enzyme gene, as well as its effectors, makes this GM strain versatile for monitoring extremely low levels of benzene in all environmental niches.

Furthermore, the public/community may see the commercial performance of biosensors produced from GM microbial strains as posing an unacceptable level of risk, reducing the government's desire for biosensor approval (Harms et al., 2006). Despite the challenges, certain biosensors now adays successfully used for *in-situ* monitoring of environmental contaminants/pollutants and commercialized (Table 2). GM microbial strains are frequently employed in MCB technology to improve the functionality of the developing biosensor towards its analyte by combining pollutant responsive genes with reporter genes (without promoter). Lindane is an organochlorine insecticide that is abundant in the environment and has negative consequences due to its neurotoxic and carcinogenic qualities. Using GM E. coli bacteria, researchers have produced a very sensitive MCB for the specific detection of lindane (Anu Prathap et al., 2012). Heavy metal cations and anions, including Cu, Fe, Mn, and Zn, are required as cofactors in a variety of metabolic pathways. However, large intracellular quantities of these metals can be harmful (Rasheed et al., 2019). The use of GM organisms for sensing and monitoring heavy metals in the environment has proven to be reliable and successful. Chen et al. (2017) characterized different accessories, such as microbial skeletal, promoters and signals, to construct a unique and sturdy microbial biosensor for analyzing detectable copper ions in a study. He also looked for fluorescent signals in betaxanthin, a plant pigment, to reduce response time. The generated GM biosensors were reported to detect copper ions selectively and sensitively in ecological samples such as tap water and pond water. Cui and his colleagues

(2018) used an *Acinetobacter baylyi* Tox2 and a luminous bacteria *Acinetobacter baylyi* ADP1 with a plasmid pWH1274 lux to investigate the cytotoxicity of seawater sample contaminated with heavy metal. Table 1 lists a variety of genetically engineered microbes that have been employed as biosensors.

5. Challenges for Microbial Biosensors

As previously stated, MCBs can be used to address a wide range of agricultural, environmental, and medical concerns. In context of sensor, work is still being done to produce new and improved MCBs, both by contribute from nature's potentiality and by developing novel genetic constructions. However, there are some obstacles in the way of broadening the use of MCBs. In remote places where handling of test samples is impractical, the development of MCBs, particularly for potent heavy metals, hazardous chemicals and promising contaminants could be extremely beneficial. Even though many microbial based biosensors developed recently, very few of them, successfully used for monitoring *in-situ* or commercialized due to technological and sociocultural obstacles that have limited their use at a larger scale. Table 3 shows the key advantages and related downsides of MCBs.

Van der Meer et al. (2004), found that no one at all developed MCBs could detect pollutants at concentrations lower than 0.1 M, and biosensors failed to respond to a group of substances (Biran et al., 2000). Furthermore, long-lasting reactions, which appear to be caused by the time necessary to express the reporter gene, could be a fundamental flaw in microbe-based biosensors for simultaneous detection. The inherent challenge of maintaining functionality of cell under poor environmental circumstances such as unavailability of nutrients or harboring inhibitory chemicals is another significant issue that has hampered MCB acceptance. Furthermore, the environmental dissemination of genetically modified microbial-derived biosensors might be harmful. Because most biosensor cells are bound by supporting materials, they are protected against unfavourable outcomes. More importantly, the development and regulation of microbe-derived biosensors may help to dispel public and community misconceptions about genetically altered microbial biosensors (Johnson et al., 2017).

Table 3. Advantages and disadvantages of microbial-derived biosensors. (Bilal and Iqbal, 2019)

Advantages		Disadvantages
√ Fast and specific detection of compounds	–	Prolonged response time
√ Concurrent monitoring of multiple compounds	–	Difficult maintenance of cell viability and activity
√ High sensitivity monitoring of bioavailable fraction of pollutant	–	Lack of durable genetic stability of engineered system
√ Cost-effective and less labor intensive than conventional sensing methods	–	Technical and societal limitations for using genetically modified strains
	–	Slow substrates and products diffusion across cell membrane into cells
	–	Influence of environmental variables (pH, temperature, nutrient availability) on biosensor functionality

6. The Future Prospects of MCBs

While we've discussed how MCBs are better suited to use in low-cost, easy operatable and handy devices, as well as several obstacles to their wider range performance, there is various exciting work worldwide that goes ahead of what was before achieved with MCBs. The future of MCBs is incredibly bright, from consumable gadgets to their integration with drone technology. In the following segment, some of the effort that is setting the path for MCB technologies in the future, as well as on the other possibilities is presented. Liu et al. provided a thorough explanation of the creation of biocompatible hydrogel elastomer hybrids five years prior using various forms of WCBs that incorporate reporter protein GFP. (2017). The researchers combined the air permeability and mechanical resilience of polydimethylsiloxane silicone elastomer with polyacrylamide (PAAm)-alginate hydrogels to generate a hybrid material that preserved cell feasibility along with excellent mechanical strength. They envision a living patch that may be applied to human skin for analyte detection of components present in blood and sweat of human, and also glove with MCBs embedded in the fingertips for monitoring of surrounds. In reaction to moisture gradients, living cells can change their form. Researchers predicted that bilayer fabricated biohybrid film may change shape and function concurrently with range of humidity, stimulated by the conformational change of living cells in varied range of moisture. E. coli MCB expressing GFP was utilized to form biohybrid multifunctional film by arranged in parallel lines to a latex surface as a proof of concept. The contractile pressure induced by cell dryness bent the film when it was exposed to dry conditions, although GFP fluorescence intensity was directly proportional to humidity. Researchers made ventilation bio flaps consisting of biohybrid material that on or off in response to sweat production during exercise and utilized it in manufacturing of prototype running suits and shoes. Above all, the incorporation of MCBs as practical components of clothing symbolises a sustainable approach that has pushed the boundaries of what MCBs were previously supposed to be capable of.

Some scientists are looking into the possibility of using MCBs in small consumable devices. Mimee and his team created an Ingestible Micro-Bio-Electronic Device (IMBED) that integrates GM MCBs with ultra-low power microelectronics to allow detection of health associated gastrointestinal biomolecules *in-situ* (Mimee et al., 2018). Other researchers have discovered the opportunity of using MCBs as probiotics for gut flora monitoring and diagnosis (Mao et al., 2018). They worked on a MCB for detecting and suppressing Vibrio cholerae. The performance of MCB with upcoming drone technology is another potential advancement in the industry. Lu et al., (2015), made a simple and handy incubation system that permits microorganisms to grow in the presence of continuous environmental monitoring in remote areas. He discusses the portable incubator system's potential applications for air pollution monitoring in remote and unexplored area. Sometime the MCBs might be incorporate with Unmanned Aerial Vehicle (UAV) and portable incubation system for simultaneous and low-cost analysis of environmental hazardous chemicals/ pollutants like carbon dioxide, ozone and nitric oxide. As the lone biological experiment, NASA recently launched the BioSentinel mission. BioSentinel's main aim was to build a biosensor that can sense and evaluate the space radiation effect on living organisms over long time periods. BioSentinel uses a biosensor that is based on the bacterium S. cerevisiae. Similar work paves the way for spore based MCBs to be used in future space missions. Future MWCBs based on fungal or bacterial spores could be created to analyze and detect toxic health related components within the spacecraft.

Finally, we see a brilliant future of MCBs which can have tremendous potential to successfully build up smart cities (Figure 4). Micro sized, independent devices installed on UAVs and maritime vehicles consisting of MCBs for the detection of a various analytes might sense and analyze water and air pollution in all counties as well as in industrial area and in agriculture area, in real-time. The huge volume of data created might be delivered remotely to central processing area, where automated computer driven programmes evaluate and understand the data, providing the valuable data of environmental quality (Moraskie et al., 2021).

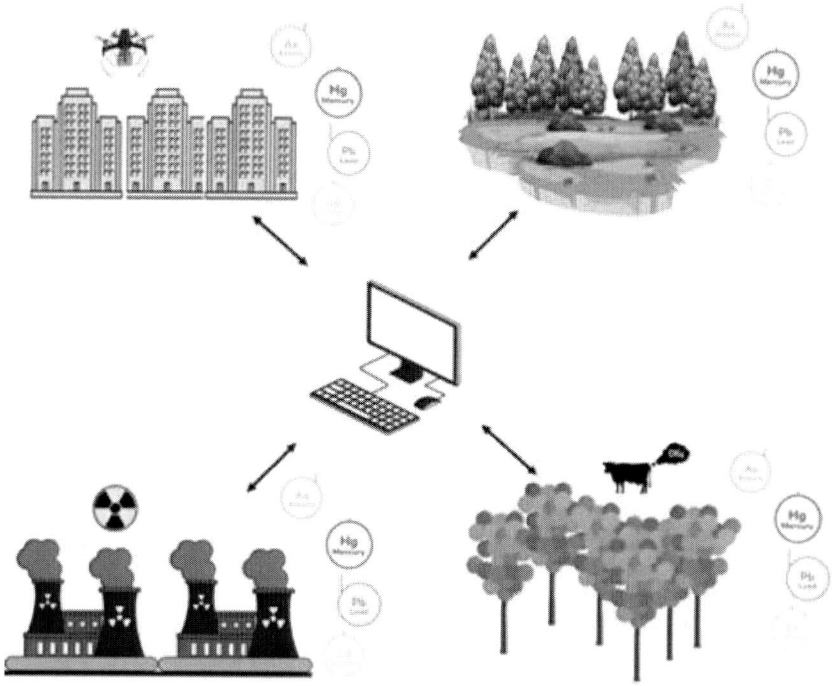

Figure 4. Future of MCBs and applicability in smart city.

Conclusion

MCBs offer an alluring replacement for their bulkier and more expensive analytical counterparts because they are a highly adaptable platform for the development of low-cost and widely available analytical devices. To measure the bioavailable levels of a target analyte or determine the toxicity of a sample, constitutive and inducible MCBs can be used. Constitutive MCBs mainly rely on observing the decline in a constitutively generated signal in response to the toxicity of a given sample, whereas inducible MCBs use genetic circuitries to produce a quantifiable signal in response to a target analyte in a dose-dependent manner. Due to the wide variety of genetic circuitries and reporters that can be utilised in MCBs, they are extremely adaptable for use with samples from a variety of scientific fields.

MCBs were developed in order to identify analytes crucial for environmental and agricultural monitoring. These comprise quorum sensing molecules, MTEs, organic compounds, and micronutrients, which are typically found in soil and water samples. Additionally, MCBs have been developed for use in biomedical applications. MCBs have advanced understanding of the microbiome in diseased condition and contributed to the creation of diagnostic tools. The field of MCB development still faces several challenges despite their laboratory successes.

Not to mention, the high adaptability and versatility of MCBs have shown tremendous promise for their potential use in novel and exciting ways. MCBs can be integrated into ingestible devices for real-time metabolite analysis of the gastrointestinal tract, into wearable clothing and living patches, and into autonomous robots and drones for the monitoring of previously inaccessible environments. As efforts to develop and optimise MCBs continue, they will become a more useful and alluring analytical tool.

References

Anu Prathap, M. U., Chaurasia, A. K., Sawant, S. N., and Apte, S. K. (2012). Polyaniline-based highly sensitive microbial biosensor for selective detection of lindane. *Analytical Chemistry*, 84, 6672–6678. doi:10.1021/ac301077d.

Ayyaru, S., and Dharmalingam, S. (2014). Enhanced response of microbial fuel cell using sulfonated poly ether ketone membrane as a biochemical oxygen demand sensor. *Analytica Chimica Acta*, 818, 15–22. doi:10.1016/j.aca.2014.01.059.

Barrios-Estrada, C., de Jesús Rostro-Alanis, M., Muñoz-Gutiérrez, B. D., Iqbal, H. M. N., Kannan, S., and Parra-Saldívar, R. (2018). Emergent contaminants: Endocrine disruptors and their laccase-assisted degradation—A review. *Sci. Total Environ.*, 612, 1516–1531.

Belkin, S. (2003). Microbial whole-cell sensing systems of environmental pollutants. *Curr. Opin. Microbiol.*, 6, 206–212.

Bilal, M., and Iqbal, H. M. N. (2019). Microbial-derived biosensors for monitoring environmental contaminants: Recent advances and future outlook, *Process Safety and Environmental Protection*, https://doi.org/10.1016/j.psep.2019.01.032.

Bilal, M., Adeel, M., Rasheed, T., Zhao, Y., and Iqbal, H. M. (2019a). Emerging contaminants of high concern and their enzyme-assisted biodegradation–A review. *Environment International*, 124, 336-353.

Bilal, M., Rasheed, T., Nabeel, F., Iqbal, H. M., and Zhao, Y. (2019b). Hazardous contaminants in the environment and their laccase-assisted degradation–A review. *Journal of environmental management*, 234, 253-264.

Bilal, M., Rasheed, T., Iqbal, H. M. N., Hu, H., Wang, W., and Zhang, X. (2018a). Horseradish peroxidase immobilization by copolymerization into cross-linked polyacrylamide gel and its dye degradation and detoxification potential. *Int. J. Biol. Macromol.*, 113, 983-990.

Bilal, M., Rasheed, T., Iqbal, H. M. N., and Yan, Y. (2018b). Peroxidases-assisted removal of environmentally-related hazardous pollutants with reference to the reaction mechanisms of industrial dyes. *Sci. Total Environ.*, 644, 1–13.

Biran, I., Babai, R., Levcov, K., Rishpon, J., and Ron, E. Z. (2000). Online and in situ monitoring of environmental pollutants: electrochemical biosensing of cadmium. *Environ. Microbiol.*, 2, 285–290.

Brunnbauer, L., Gonzalez, J., Lohninger, H., Bode, J., Vogt, C., Nelhiebel, M., Larisegger, S., and Limbeck, A. (2021). Strategies for trace metal quantification in polymer samples with an unknown matrix using Laser-Induced Breakdown Spectroscopy. Spectrochim. *Acta Part B At. Spectrosc.*, 183, 106272.

Chan, A. C., Ager, D., and Thompson, I. P. (2013). Resolving the mechanism of bacterial inhibition by plant secondary metabolites employing a combination of whole-cell biosensors. *Journal of Microbiological Methods*, 93, 209–217. doi:10.1016/j.mimet.2013.03.021.

Chang, H. J., Voyvodic, P. L., Zúñiga, A., and Bonnet, J. (2017). Microbially derived biosensors for diagnosis, monitoring and epidemiology. *Microb. Biotechnol.*, 10(5), 1031-1035.

Cui, Z., Luan, X., Jiang, H., Li, Q., Xu, G., Sun, C., and Huang, W. E. (2018). Application of a bacterial whole cell biosensor for the rapid detection of cytotoxicity in heavy metal contaminated seawater. *Chemosphere*, 200, 322-329.

Daunert, S., Barrett, G., Feliciano, J. S., Shetty, R. S., Shrestha, S., and Smithspencer, W. (2000). Genetically engineered whole-cell sensing systems: Coupling biological recognition with reporter genes. *Cheminform.*, 100, 2705–2738.

Deng, F., Zhang, D., Yang, L., Li, L., Lu, Y., Wang, J., Fan, Y., Zhu, Y., Li, X., and Zhang, Y. (2021). Effects of antibiotics and heavy metals on denitrification in shallow eutrophic lakes. *Chemosphere*, 291, 132948.

Dhote, S. S., Deshmukh, L., and Paliwal, L. (2013). Miceller chromatographic method for the separation of heavy metal ions and spectrophotometric estimation of UO2 2+ on bismuth silicate layer. *Int. J. Chem. Azal. Sci.*, 4, 85–90.

Di Lorenzo, M., Thomson, A. R., Schneider, K., Cameron, P. J., and Ieropoulos, I. (2014). A smallscale air-cathode microbial fuel cell for on-line monitoring of water quality. *Biosensors and Bioelectronics*, 62, 182–188. doi:10.1016/j.bios.2014.06.050.

Diaz, E., and Prieto, M. A. (2000). Bacterial promoters triggering biodegradation of aromatic pollutants. *Curr. Opin. Biotech.*, 11, 467–475.

Du, Z., Li, H., and Gu, T. (2007). A state of the art review on microbial fuel cells: A promising technology for wastewater treatment and bioenergy. *Biotechnology Advances*, 25, 464–482. doi:10.1016/j.biotechadv.2007.05.004.

Du, H., Strohsahl, C. M., Camera, J., Miller, B. L., and Krauss, T. D. (2005). Sensitivity and specificity of metal surface-immobilized “molecular beacon” biosensors. *J. Am. Chem. Soc.*, 127, 7932–7940.

Gavrilas, S., Ursachi, C. S., Pert, a-Cris, S., and Munteanu, F. D. (2022). Recent Trends in Biosensors for Environmental Quality Monitoring. *Sensors*, 22, 1513. https://doi.org/10.3390/s22041513

Gennaro, P., Bruzzese, N., Anderlini, D., Aiossa, M., Papacchini, M., Campanella, L., and Bestetti, G. (2011). Development of microbial engineered whole-cell systems for environmental benzene determination. *Ecotox. Environ. Safe.*, 74(3), 542-549.

Gui, Q., Lawson, T., Shan, S., Yan, L., and Liu, Y. (2017). The application of whole cell based biosensors for use in environmental analysis and in medical diagnostics. *Sensors*, 17(7), 1623.

Hansen, L. H., and Sørensen, S. J. (2001). The use of whole-cell biosensors to detect and quantify compounds or conditions affecting biological systems. *Microb. Ecol.*, 42, 483–494.

Harms, H., Wells, M. C., and van der Meer, J. R. (2006). Whole-cell living biosensors—are they ready for environmental application? *Appl. Microbiol. Biotechnol.*, 70, 273–280.

Hernandez-Vargas, G., Sosa-Hernández, J. E., Saldarriaga-Hernandez, S., VillalbaRodríguez, A. M., Parra-Saldivar, R., and Iqbal, H. M. N. (2018). Electrochemical biosensors: A solution to pollution detection with reference to environmental contaminants. *Biosensors*, 8(2), 29.

Johnson, A. O., Gonzalez-Villanueva, M., Wong, L., Steinbüchel, A., Tee, K. L., Xu, P., and Wong, T. S. (2017). Design and application of genetically-encoded malonyl-CoA biosensors for metabolic engineering of microbial cell factories. *Metabol. Eng.*, 44, 253-264.

Keane, A., Phoenix, P., Ghoshal, S., and Lau, P. C. K. (2002). Exposing culprit organic pollutants: a review. *J. Microbiol. Meth.*, 49, 103–119.

Kim, M., Lim, J. W., Kim, H. J., Lee, S. K., Lee, J. L., and Kim, T. (2015). Chemostat-like microfluidic platform for highly sensitive detection of heavy metal ions using microbial biosensors. *Biosensors and Bioelectronics*, 65, 257–264. doi:10.1016/j. bios.2014.10.028.

Kumar, H., Kumari, N., and Sharma, R. (2020). Nanocomposites (conducting polymer and nanoparticles) based electrochemical biosensor for the detection of environment pollutant: Its issues and challenges. *Environ. Impact Assess. Rev.*, 85.

Li, L., He, J., Gan, Z., and Yang, P. (2021). Occurrence and fate of antibiotics and heavy metals in sewage treatment plants and risk assessment of reclaimed water in Chengdu, China. *Chemosphere*, 272, 129730.

Liu B., Lei Y., and Li B. (2014). A batch-mode cube microbial fuel cell based "shock" biosensor for waste water quality monitoring. *Biosensors and Bioelectronics*, 62, 308–314. doi:10.1016/j.bios.2014.06.051.

Liu, Q., Wu, C., Cai, H., Hu, N., Zhou, J., and Wang, P. (2014). Cell-based biosensors and their application in biomedicine. *Chem. Rev.*, 114, 6423–6461.

Looger, L. L., Dwyer, M. A., Smith, J. J., and Hellinga, H. W. (2003). Computational design of receptor and sensor proteins with novel functions. *Nature*, 423, 185–190.

Lu, Y., Macias, D., Dean, Z. S., Kreger, N. R., and Wong, P. K. (2015). IEEE Trans. *NanoBioscience*, 14 (8), 811–817.

Maduraiveeran, G., and Jin, W. (2017). Nanomaterilas based electrochemical sensor and biosensor platforms for environmental applications. *Trends Environ. Anal. Chem.*, 13, 10–23.

Mao, N., Cubillos-Ruiz, A., Cameron, D. E., and Collins, J. J. (2018). *Sci. Transl. Med.*, 10 (445), eaao2586.

Mimee, M., Nadeau, P., Hayward, A., Carim, S., Flanagan, S., Jerger, L., Collins, J., McDonnell, S., Swartwout, R., Citorik, R. J., Bulovi´c, V., Langer, R., Traverso, G., Chandrakasan, A. P., and Lu, T. K. (2018). An ingestible bacterial-electronic system to monitor gastrointestinal health, *Science*, 360 (6391), 915.

Moraskie, M., Roshid, M. H. O., O'Connor, G., Dikici, E., Zingg, J. M., Deo, S., and Daunert, S. (2021). Microbial whole-cell biosensors: Current applications, challenges, and future perspectives. *Biosens. Bioelectron.*, 191, 113359.

Mulchandani A., Rajesh. (2011). Microbial biosensors for organophosphate pesticides. *Applied biochemistry and biotechnology*, 165, 687–699. doi:10.1007/s12010-0119288-x.

Murzyn, C. M., Allen, D. J., Baca, A. N., Ching, M. L., and Marinis, R. T. (2021). Tunable Infrared Laser Absorption Spectroscopy of Aluminum Monoxide A2Πi-X2Σ. *J. Quant. Spectrosc. Radiat. Transf.*, 279, 108029.

Neethirajan, S., Ragavan, V., Weng, X., and Chand, R. (2018). Biosensors for Sustainable Food Engineering: Challenges and Perspectives. *Biosensors*, 8(1), 23.

Niazi J. H., Kim B. C., Ahn J. M., and Gu M. B. (2008). A novel bioluminescent bacterial biosensor using the highly specific oxidative stress-inducible pgi gene. *Biosensors and Bioelectronics*, 24, 670–675. doi:10.1016/j.bios.2008.06.026.

Nigam, V. K., and Shukla, P. (2015). Enzyme Based Biosensors for Detection of Environmental Pollutants—A Review. *J. Microbiol. Biotechnol.*, 25, 1773–1781.

Peng, H., Hou, B., Huana, Q., Fan, Y., Bilal, M., Wang, W., and Bennett, G. N. (2018). Oxidative photo-catalyzed degradation of a new biological fungicide, phenazine-1carboxylic acid. Desalin. *Water Treat.*, 115, 207-212.

Rajkumar, P., Ramprasath, T., and Selvam, G. S. (2017). A simple whole cell microbial biosensors to monitor soil pollution. *New Pesticides and Soil Sensors*, 437–481.

Rasheed, T., Bilal, M., Nabeel, F., Adeel, M., and Iqbal, H. M. (2019). Environmentally related contaminants of high concern: Potential sources and analytical modalities for detection, quantification, and treatment. *Environ. Int.*, 122, 52–66.

Rasheed, T., Bilal, M., Nabeel, F., Iqbal, H. M. N., Li, C., and Zhou, Y. (2018a). Fluorescent sensor based models for the detection of environmentally-related toxic heavy metals. *Sci. Total Environ.*, 615, 476–485.

Rasheed, T., Li, C., Bilal, M., Yu, C., and Iqbal, H. M. (2018b). Potentially toxic elements and environmentally-related pollutants recognition using colorimetric and ratiometric fluorescent probes. *Sci. Total Environ.*, 640, 174–193.

Raut, N., O'Connor, G., Pasini, P., and Daunert, S. (2012). Engineered cells as biosensing systems in biomedical analysis. *Anal. Bioanal. Chem.*, 402, 3147–3159.

Ron, E. Z. (2007). Biosensing environmental pollution. *Curr. Opin. Biotech.*, 18, 252– 256.

Shen, Y., Wang, M., Chang, S., and Ng, H. Y. (2013). Effect of shear rate on the response of microbial fuel cell toxicity sensor to Cu(II). *Bioresource Technology*, 136, 707–710. doi:10.1016/j.biortech.2013.02.069.

Shin, H. J. (2010). Development of highly-sensitive microbial biosensors by mutation of the nahR regulatory gene. *Journal of Biotechnology*, 150, 246–250. doi:10.1016/j.jbiotec.2010.09.936.

Sochor, J., Zitka, O., Hynek, D., Jilkova, E., Krejcova, L., Trnkova, L., and Kizek, R. (2011). Bio-sensing of cadmium (II) ions using Staphylococcus aureus. *Sensors*, 11(11), 10638-10663.

Tian, Y., Lu, Y., Xu, X., Wang, C., Zhou, T., and Li, X. (2017). Construction and comparison of yeast whole-cell biosensors regulated by two RAD54 promoters capable of detecting genotoxic compounds. *Toxicol. Mech. Meth.*, 27, 115–120.

Trapananti, A., Eisenmann, T., Giuli, G., Mueller, F., Moretti, A., Passerini, S., and Bresser, D. (2021). Isovalent vs. aliovalent transition metal doping of zinc oxide lithium-ion battery anodes—In-depth investigation by ex situ and operando X-ray absorption spectroscopy. *Mater. Today Chem.*, 20, 100478.

van der Meer, J. R., Tropel, D., and Jaspers, M. (2004). Illuminating the detection chain of bacterial bioreporters. *Environ. Microbiol.*, 6, 1005–1020.

Vollmer, A. C., and Van Dyk, T. K. (2004). Stress responsive bacteria: biosensors as environmental monitors. *Adv. Microb. Physiol.*, 49, 131–174.

Wang, X., Liu, M., Wang, X., Wu, Z., Yang, L., Xia, S., Chen, L., and Zhao, J. (2013). Pbenzoquinone-mediated amperometric biosensor developed with Psychrobacter sp. for toxicity testing of heavy metals. *Biosensors and Bioelectronics*, 41, 557–562. doi:10.1016/j.bios.2012.09.020.

Wang, X., Lu, X., and Chen, J. (2014). Development of biosensor technologies for analysis of environmental contaminants. *Trends Environ. Anal. Chem.*, 2, 25-32.

Wei, T., Zhang, C., Xu, X., Hanna, M., Zhang, X., Wang, Y., Dai, H., and Xiao, W. (2013). Construction and evaluation of two biosensors based on yeast transcriptional response to genotoxic chemicals. *Biosensors and Bioelectronics*, 44, 138–145. doi:10.1016/j.bios.2013.01.029.

Wu, W., Qu, S., Nel, W., and Ji, J. (2022). Tracing and quantifying the sources of heavy metals in the upper and middle reaches of the Pearl River Basin: New insights from Sr-Nd-Pb multi-isotopic systems. *Chemosphere*, 288, 132630.

Xiong, J., Sun, Z., Yu, J. H., Liu, H., and Wang, X. D. (2022). Thermal self-regulatory smart biosensor based on horseradish peroxidase immobilized phase-change microcapsules for enhancing detection of hazardous substances. *Chem. Eng. J.*, 430, 132982.

Xu, P. (2018). Production of chemicals using dynamic control of metabolic fluxes. *Curr. Opin. Biotechnol.*, 53, 12-19.

Xu, X., and Ying, Y. (2011). Microbial biosensors for environmental monitoring and food analysis. *Food Rev. Int.*, 27(3), 300-329.

Yagi, K. (2007). Applications of whole-cell bacterial sensors in biotechnology and environmental science. *Appl. Microbiol. Biotechnol.*, 73, 1251–1258.

Chapter 3

Recent Trends in the Detection of Staphylococcal Enterotoxins in the Food Matrix

Smriti Singh
and Seema Nara*
Department of Biotechnology, Motilal Nehru National Institute of Technology Allahabad, Prayagraj, Uttar Pradesh, India

Abstract

Staphylococcal food poisoning (SFP) is one of the most frequent foodborne outbreaks worldwide. SFP occurs due to ingestion of Staphylococcal extracellular toxins/enterotoxins (SEs) in contaminated food. These enterotoxins are released by Staphylococcus aureus, display super-antigenic characteristics and a very small amount (~1µg/gm food) can be sufficient to cause food poisoning, and severe infections in immune-compromised persons and children. To date only ELISA-based conventional methods are used for the detection of SEs in food matrices. These techniques suffer from limitations of low sensitivity, high cost, time consuming and possible denaturation of enzymes and antibodies under harsh conditions. To overcome these issues, there is a need of developing more sensitive, cost-effective, fast and user-friendly new techniques for the SE's detection in food. Aptamers along with nanozymes can be a better alternative in the detection of SEs. These entities enhance the sensitivity and specificity of detection to greater extent. This chapter focuses on SEs and its families, and on new sensing approaches based on nanozymes and aptamers for the detection of SEs in

* Corresponding Author's Email: seemanara@mnnit.ac.in, seemanara@gmail.com.

In: Biosensing
Editor: Rushika Patel
ISBN: 979-8-88697-911-4

food products. The role of Point-of- care-testing (POCT) and multiplex techniques in the detection of SEs are also discussed.

Keywords: aptasensors, nanobiosensors, point-care-of-testing, staphylococcal enterotoxins, staphylococcal food poisoning

1. Introduction

Food safety can only be assured by the early and timely detection of contaminants, contagious microorganisms and their toxins existing in food products. Food poisoning or food-borne illness occurs by consumption of intoxicated food infected with microbial toxins (single or multiple types toxins) (Argudin MA, Mendoza MC, Rodicio MR, 2010). These microbial toxins are as well recognized as superantigens (SAs) and lead to the activation of very strong immune response (Kotb M, Fraser JD, 2008). Food poisoning caused by ingestion of enterotoxins in contaminated food led to severe and sometimes mortal diseases, making it a public health issue worldwide (Nunes MM, Caldas ED 2017, Wunderlichova L et al., 2014, Scallan E et al., 2011). Staphylococcal enterotoxins (SEs) secreted by *Staphylococcus aureus* strains, are highly stable and remain active in food products even after heat treatment like pasteurization (Tatini, 1976; Balaban and Rasooly, 2000; Larkin et al., 2009; Ding et al., 2016). SEs are proteins having a molecular weight of 27-34 kDa, 21 different types, unassailable by a gastrointestinal enzyme (pepsin and trypsin) (Seo JH et al., 2017). Out of these 21 SEs, SEA, SEB, SEC, SED and SEE are most common and accountable for staphylococcal food poisoning (SFP) (Hennekine MJ, De S, Byser D, 2012). SEA and SEB are predominantly reported in most of SFP cases in many countries (Hennekine MJ, De S, Byser D, 2012, Argudin, MA, Mendoza MC, Rodicio MR, 2010, Heidinger JC, Winter CK, 2000). Intoxication of food with SEs induces symptoms like vomiting, emesis, abdominal cramps, nausea, and diarrhea. Roughly 100 ng of SEA and 20 ng of SEB toxin are enough to cause SFP symptoms in adults (Carfora V et al., 2015, Marrack P, Kappler J, 1990, Evenson ML et al., 1988).

Most traditional methods for SEs detection in food are culture-based techniques. Secretion of toxins depends upon the culture conditions like incubation time, pH, media etc. (Chatrathi MP, Wang J and Collins GE, 2007). With time, researchers have developed various methods for detecting SEs with better sensitivity and specificity, including polymerase chain reaction (PCR)-based assays (Fusco V et al., 2011), sandwich enzyme-linked immunosorbent assay (ELISA) (Nagraj S et al., 2016), various biosensors (Tallent SM et al., 2013, Pinheiro L et al., 2015, Chang L, Li JM, Wang LN, 2016), while RPLA (reversed passive latex agglutination tests) has quite low sensitivity and specificity (Table 1). These methods are selective and reproducible, but require sophisticated apparatus, well skilled personnel, long analysis time, and has limited sensitivity (Tang D et al., 2010, Lin HC, Tsai WC 2003, Campbell GA, Medima MB, Mutharasan R, 2007, Yang M et al., 2015, Zhu S et al., 2009, Yang M et al., 2009, Sospendra I et al., 2012 (a), Sospendra I et al., 2012(b)).

The sensitivity of these techniques can be enhanced by the coupling of nanoparticles as signal enhancers. In the last few decades, researchers developed assays with good sensitivity by using nanoparticles as a signal enhancer and also an alternative to enzymes. Although the use of antibodies as bio-recognition molecules, conventional detection methods have some limitations and disadvantages, like high production cost, time consumption, and batch variation in production. Antibodies can be inactivated with harsh treatment and modifications, even their affinity and specificity also depend upon the environmental conditions, making them unfavorable for food and environmental analysis (Mondal B et al., 2015). To address the limitations of antibodies, aptamers have emerged as alternatives to antibodies as recognizing elements. Easily functionalization, high stability to harsh treatment, high purity, an extensive variety of targets, and low-cost preparation are some excellent properties of aptamers (Blind M and Blank M, 2015). These short sequences are acquired through the SELEX (Systematic evolution of ligands by exponential enrichment) process and have a broad range of targets (Stoltenburg R, Reinemann C and Strehlitz, 2007). The sensing devices can be designed with increased sensitivity, less detection time, and enable multiplexing capability by exploiting the unique properties of nanoscale materials (Farka Z et al., 2017). In the last few decades, research in nanotechnology has offered technological advancement in the detection of foodborne disease-causing pathogens, toxins and other contaminants (Leonard P et al., 2003, Valdes MG et al., 2009, Lopez BP, Merkoci A, 2011, Ali ME et al., 2011 (a), Ali ME et al., 2011 (b), Ali ME et al., 2011(c), Ali ME et al., 2012, Sonawane SK et al., 2014, Kaittanis C et al., 2016). Aptamer-based biosensors can be integrated with nanoparticles to enhance the sensitivity and selectivity of technique for SEs detection in food matrices.

Table 1. Commercially available kits for the detection of SEs in food products

SE	Platform available	Method	LOD	Food sample	Limitations
	VIDAS® SET (BioMerieux, Europe USA)	*ELFA		Dairy products, meat, and seafood, etc.	1. Sample matrix interference 2. Requires extraction of SE from food matrices. 3. Time of analysis: 80-90 minutes 4.Crossreactivity with A/E, E/A, B/C, and C/B
A-E	VIDAS® SET II (BioMerieux, Europe USA)		0.25 ng/gm	Ready to eat, raw milk, processed food, fruit juices, canned food etc.	
A-E and G-I	VIDAS® SET III (BioMerieux, Europe USA)		2.09 ng/mL	Milk, custard, chocolate pudding, Rioctta cheese, Egg salad, smoked Salmon	
A-/E	SET RPLA (Denka Seiken, Japan)	*RPLA	0.5 ng/gm	Ready to eat, raw milk, processed food, fruit juices, canned food etc.	
A-E	TRANSIA® PLATE SET	*ELISA	0.2 ng/gm	Ready to eat, raw milk, processed food, fruit juices, canned food etc.	
A-E	TECRA SE VIA (3M Microbiology Canned)	*ELISA	1 ng/mL	Canned mushrooms, non-fat dry milk, canned lobster bisque, beef, and pasta, cooked chicken, and cheese	
A-E	RIDASCREEN I R - Biopharm, (Darmstadt, Germany)	*ELISA	0.1 mg/mL	Ice-cream, milk and dairy products, pastries, cake stuffing, egg products, salads, pasta, ham, pies, finished meat products, chicken meat products, fish, and fish products	
A-D	SET- RPLA (Oxoid)	RPLA	1 ng/mL	A broad range of food and food products like dairy products, meat, and meat products	

*ELFA: Enzyme-linked fluorescent immunoassay, *ELISA: Enzyme-linked fluorescent immunoassay, *RPLA: Reverse Passive Latex Agglutination, *SE: Staphylococcal Enterotoxin

2. Emerging Technologies in Staphylococcal Enterotoxin Detection

To address the drawbacks of existing conventional detection methods, attention is paid to developing new techniques for the detection of food toxins. Immunosensors are instruments/devices that can recognize biological entities by their specific interaction with recognizing/capturing molecules which can be either antibodies/antigen or aptamer (DNA/RNA/peptide). The biological interaction can be monitored either through generated optical (color, fluorescence) or electrical signals whose intensity is proportional to the concentration of the analyte (Kasai S et al., 2000). Nanoparticles can be used as label/signal enhancer/enzyme alternatives to significantly enhance the sensitivity of detection techniques with both bio-recognition molecules (antibodies and aptamer). Carbon nanotubes (CNTs), gold nanostructures (GNS), reduced graphene oxide (rGO), and quantum dots (QDs) are widely used in nanobiosensors devices. POCT devices are also developed by allowing miniaturization and simplification of testing devices with good sensitivity and specificity, as these are accessible and can be used by nonprofessional persons. This section presents some emerging detection methods using nanoparticles and aptamer for fast, reliable, sensitive, specific detection of SEs.

2.1. Optical Sensing Methods

Optical immunosensors estimate the change in absorption/fluorescence/ luminescence/scattering pattern of light upon passing the reaction medium for the purpose of analyte detection. These methods offer sensitive, fast, and real-time monitoring of SEs. Yang M and coworkers (2008) developed fluorescence-based immunosensor using CNTs for SEB detection as CNTs provide a large surface area to load more antibody molecules and allow sensitive detection. The sandwich immunosensor generated a six to eight-time higher signal than of standard type antibody-antigen ELISA assay, with a linear detection range of 100-0.1 ng/mL with the LOD of 0.1 ng/mL (Yang M et al., 2008). Fluorescence based detection methods required a light excitation system for fluorophore along with a complex optical system of lens and filter system for detection. Whereas there is no requirement of such complex systems with enhanced chemiluminescence (ECL) as it emits lights directly and can be easily detected by simple instrumentation. The ECL improves signal generation from enzymatic reactions (Horseradish Peroxidase (HRP), Alkaline phosphatase, Glucose oxidase (GO)), by using chemiluminescence moieties such as isoluminol, luminol or nanoparticles (metal-based labels) as enhancers. Enzymes catalyze substrate, whereas enhancers produce excited intermediates that can be detected by simple instruments. Yang et al., (2009) exploited GNPs as enhancers for developing ECL immunosensor for the detection of enterotoxins in food (Yang M et al., 2009). For detection purposes, sandwich of HRP-conjugated reporting antibody (anti rabbit IgG), SEB and GNPs-conjugated capturing antibody was developed for ELISA. The LOD for SEB was 0.01 ng/mL, and the assay was 10 times more sensitive than ELISA and CNT-based biosensors (Yang M et al., 2008, 2009). Although CNTs were also required functionalization and shortening and less biocompatible. While GNPs easily synthesize and are less toxic in nature, so GNPs based nanobiosensor can be easily prepared. Later, Haddada MB et al., 2017 employed LSPR shift-mediated color change of GNPs for developing a sensitive immunosensor detecting SEA in milk samples. Anti-SEA Ab was treated with Traut's reagent to introduce thiol modification for stable conjugation with GNPs. In presence of enterotoxin type-A, GNPs-anti SEA Abs and SEA formed an immuno-complex that led to the aggregation of GNPs. An increase in the nanoparticle size leads to an LSPR shift from red to blue color. This direct assay had 5 ng/mL of LOD for SEA in milk products (Haddada MB et al., 2017). Similarly, Zhou D et al., 2017 designed an aptasensor with DNAzyme to detect SEB in the real samples. The assay employed G-quadruplex as HRP alternative and a change in LSPR of GNPs was used as an indicator. G-quadruplex (DNAzymes) is a hemin aptamer possessing HRP-like activity in presence of hemin. In the absence of SEB, G-quadruplex aptamer partially hybridizes with SEB-specific aptamer to form a DNA duplex. The added H_2O_2 rapidly reduced Au ions to generate a quasi-spherical shape GNPs solution of red color. While in the presence of the target (SEB), SEB and aptamer interacted with each other and led to a conformational change in DNA duplex. This resulted in the liberation of G-quadruplex forming sequences. Upon addition of hemin, G-quadruplex DNAzyme hydrolyzed the H_2O_2, and the reduction of Au ions become eventually slow. This leads to aggregation of GNPs and SPR shift from red to blue. This assay was reported to have a linear range of 0.1 to 500 pg/mL and LOD of 1 pg/mL for SEB detection (Zhou D et al., 2017). Nanozyme catalyzes substrate only when there is no surface masking and their surface is accessible for the substrate. Tan F et al., 2018 and Xie X et al., 2019 developed a colorimetric aptamer-based biosensor for detecting SEB by exploiting the masking effect on eggshell membrane-templated SH-Gold Nanoclusters (GNCs) and SH-GNCs-chitosan respectively. Here H_2O_2 mediated reduction of $HAuCl_4$ to form wine red color GNPs was used as a colored endpoint signal for monitoring the reaction. The underlying principle was that the GNC-aptamer surface was masked in the presence of target SEB and prevented H_2O_2 from catalytic decomposition. The H_2O_2 was then available to reduce $HAuCl_4$ and produce wine red color solution. However, in the absence or low concentration of target SEB, the GNCs surface is unmasked and decomposed H_2O_2. Here with a limited amount of H_2O_2 available, it resulted in aggregation of GNP and solution color change from red to violet. Therefore, a change in the tone of the solution was used as a signal to sense the SEB concentration (as shown in Figure 1). These colorimetric assays were reported to have a LOD of 0.12 ng/mL for GNCs-ESM (Tan F et al., 2018) and 1 pg/mL for GNCs-Chitosan (Xie X et al., 2019).

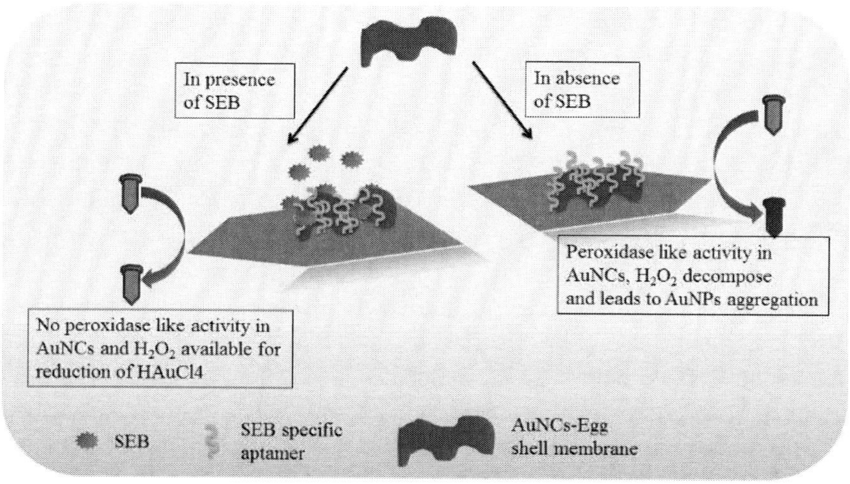

Figure 1. Schematic representation of H_2O_2 mediated GNPs synthesis as signal for SEB detection. (Ref. Tan F et al. 2018)

Mondal B et al., 2018 developed a label-free colorimetric nanobiosensor for SEB by exploiting target salt-based aggregation of GNPs and aptamer affinity towards their targets. SEB specific aptamer binds with citrate capped GNPs, GNPs remain dispersed even after addition of high concentration of NaCl. While in the presence of SEB, target induced conformational change in stem-loop of apatmer and this eventually led to the aptamer release from the surface of GNPs and aggregation of GNPs. In this colorimetric assay, 50 ng/mL of SEB can be detected visually and 0.5 ng/mL via spectroscopic method (Mondal B et al., 2018). Guanine and cytosine rich fluorescent DNA sequences was used as template to synthesize silver nanoclusters (AgNCs) with enhanced fluorescence intensity of DNA molecule (Guo et al., 2019, Zhu J et al., 2015). The DNA-AgNCs were used as fluorescent probe and poly pyrole nanoparticles (PPyNPs) as quencher by Zhang et al., 2020 to develop an aptasensor for sensitive detection of SEA. The 5'end (C6G5C6AgNCs) of the probe possess fluorescence and its 3' end sequence was complementary to bind SEA specific aptamer. In the absence of SEA, aptamer-DNA-AgNCs adsorb on PPyNPs surface through powerful π- π stacking interconnections and subsequent energy transfer take place from AgNCs (act as energy donor) to PPyNPs (act as energy acceptor) resulted in quenching of fluorescence of DNA probe. While in the presence of SEA, aptamer-DNA-AgNCs desorbed from PPyNPs surface to form high affinity aptamer-SEA complex. This eventually resulted in the fluorescence recovery of DNA-AgNCs, which was linearly related with SEA concentration. The Lowest limit of detection for SEA was 3.393 μg/mL and in linear detection range of 0.5 -1×10^{-3} μg/mL (Zhang X et al., 2020).

Aptamer performance and sensitivity can be significantly improved in different colorimetric (Lu SS et al., 2016, Liu P et al., 2013), fluorescence (Sharma A et al., 2015, Blind M and Blank M 2015), chemiluminescence (Wang H et al., 2014) and electrochemical (Zhuang J et al., 2013, Liu M et al., 2017, Zhao J et al., 2012) biosensing system, by using sequential enzyme-free nucleic acid (NA) amplification. Hybridization chain reaction (HCR) (Dirks RM and Pierce NA, 2004) is one of the enzyme-free NA amplification technique(s). Two hairpin DNA molecules having long repeating units can form a DNA nanostructure, when toehold-mediated strand displacement (TMSD) occurred in presence of an initiator DNA molecule (Zhang DY and Seelig G, 2011). Briefly, TMSD process starts with a dsDNA complex composed of the original strand with overhanging region (Toehold) and the protector strand (Guo Y et al., 2017). Toehold region having complementarity to a third DNA strand referring as the "invading/initiator strand" (Guo Y et al., 2017, Zhang D and Seelig G 2011). Because of having complementary regions, the hybridization process starts between toehold region and initiator DNA strand with progressive TMSD and a DNA complex of three DNA strands was formed (Guo Y et al., 2017, Yurke B and Millis AP, 2003). With this HCR concept, Xu Y et al., 2018 designed a green fluorescence tagged aptasensor for the detection of SEB in milk samples. In this assay, two hairpin DNA probes, first one (H1) tagged with a fluorophore (6-FAM) at 5' end and quencher (BHQ-1) at 3' end, and another (H2) with an initiator DNA (cDNA) partially hybridized with SEB specific aptamer. In the absence of SEB, aptamer temporarily hybridized with cDNA and HCR amplification could not take place, and probes remained in folded form. In this hairpin form, 6-FAM and 1-BHQ were in close proximity; therefore, the fluorescence of 6-FAM was quenched by BHQ-1 and generated a low signal. While in the presence of target, aptamer dissociated from cDNA (initiator DNA) and formed an aptamer-target complex. Then, cDNA triggered the unfolding of probes leading to hybridization between probes and causing 6-FAM to fluoresce. The assay was reported to have a detection limit as low as 3.13 ng/mL with a dynamic range of detection of 3.13 ng/mL to 100 ng/mL (Xu Y et al., 2018).

2.2. Lateral Flow Immunoassay (LFIA)

LFIAs are most promising and simple tool; to identify a variety of analytes, pollutants and pathogens at the POCT/near to site/at home. In LFIAs, movement of liquid sample is facilitated via capillary action along the paper or membrane-based cassette system. These immunochromatography strips offer rapid, cost-effective, and user-friendly mode of operation in comparison to lab set-up. In LFIA setup, membranes like sample pad, Nitrocellulose membrane (test pad), and adsorption pad/reservoir overlap to facilitate the migration of the target with running buffer. Reporter moieties (such as enzymes, fluorophores or nanoparticles) are used for the confirmation of the presence or absence of analytes in form of visible color change at the test line and control line. Recently nanoparticles gained a lot of attraction as reporter moieties because of their excellent mechanical, catalytic, electrical, and optical properties contribute to enhancing the sensitivity and specificity of LFIA strips. Colored nanostructures such as GNPs (Odom TW et al., 1998, Qu B et al., 2008)), AgNPs (Wu KH et al., 2019), Carbon (Yang MH et al., 2010), magnetic nanoparticles MNPs (Kim MS et al., 2017), and QDs (Zhao Y et al., 2017) are used as label or reporter moieties for antibody in the detection of target analytes.

First LFIA strip for staphylococcal enterotoxin detection was developed by Rong-Hwa et al., 2010 for the detection of SEB with GNPs conjugated anti-species polyclonal antibodies (. The presence and absence of red color at T-line showed positive and negative results respectively. Lateral flow assay (LFA) strip detected SEB as low as1 ng/mL within 5 minutes without any cross reactivity with SEs (B, E, C, D and E). The sensitivity of the assay was further enhanced by silver staining to achieve a LOD up to 10 pg/mL (Rong-Hwa S, et al., 2010). Polyclonal antibodies may show more cross-reactivity, affinity and batch-to-batch variation; hence these demerits limit their application as capturing/recognizing moieties. Unlike polyclonal Abs, Monoclonal Abs having high homogeneity, low background noise, and less cross reactivity. Nara and her group (2018) developed LFA strip based on the formation of antibody-antigen immuno-complex with GNPs as label moieties for detection of SEA in milk sample (as shown in Figure 2). The LOD of this LFIA strip was reported to be 0.5 μg/mL (Upadhyay N and Nara S, 2018).

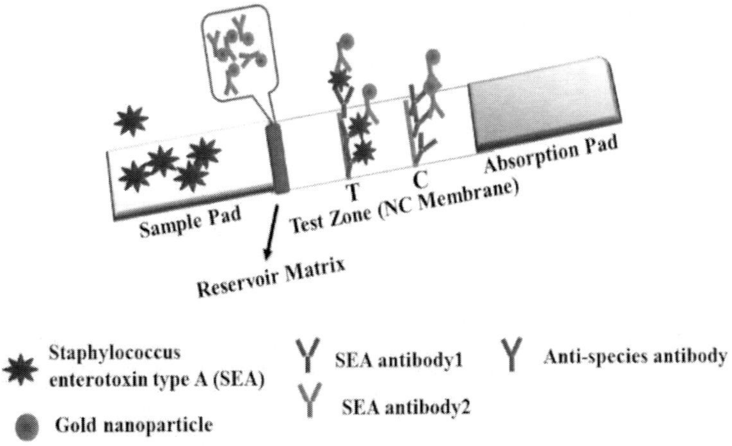

Figure 2. Schematic representation of sandwich LFIA for SEA detection (Adopted and reprint permission from Nara et al., 2018).

Similarly, Wu K-H et al., (2020) designed LFIA for SEB detection by using citrate stabilized AgNPs with LOD of 0.5 ppm in food samples. A brown color developed within 15 minutes at both Test and Control lines in the presence of SE in food samples. This sandwich type LFIA showed no cross-reactivity with homologous SE (Wu K-H et al., 2020).

2.3. Electrochemical Sensing

An electrochemical (EC) biosensor is a device that can estimate and determine the target by generating EC signals, which are proportional to the amount of target molecules (Thipmanee O et al., 2016, Xuan F, Fan TW and Hsing I-M, 2015). EC transducer and bio-probe are two main components of EC biosensor, where the transducer converts the biological signal into EC signal, and bioprobe is the recognition element. The sensitivity and selectivity of the EC sensing system depend upon the binding affinity of the target and bio-recognition probe. In the last decade, researchers paid more attention towards developing electrochemical biosensors for the detection of food-borne pathogens and their toxins because of their simplicity, low power requirements, high sensitivity, and specificity with ease of modification

(Braiek M et al., 2012, Tran QH et al., 2012). Tang et al., developed an electrochemical nanobiosensor by using HRPSiCNTs (HRP-nanosilica-dopped multiwalled CNTs) being a signal amplifier for SEB detection. A sandwich-type immuno-complex formed between anti-SEB-labelled HRPSiCNTs and immobilized anti-SEB on screen printed carbon electrode (SPE) in the presence of SEB. With the help of thionine, HRP catalyzed H_2O_2 and led to increase in reduction current. This increase in current is proportional to SEB concentration in samples. The large surface area of CNTs allowed the loading of more HRP molecules on its surface, which finally resulted in the amplification of an electrochemical signal. This immunosensor reported LOD of 10 pg/mL with a linear range of 0.05-15 ng/mL for SEB. It also provided more sensitive and accurate detection of SEB compared to ELISA with spiked food matrices like juice, soymilk, pork food, watermelon, and apple juice (Tang et al., 2010). Sharma et al., designed a sandwich electrochemical nanobiosensor for enterotoxin B detection by immobilizing polyclonal SEB-specific antibodies on screen printed electrode (SPE) and using Monoclonal Ab conjugated with PbS-QDs/ZnS-QDs as a probe. With the addition of 1N HCl solution, trapped QDs dissolved to release Pb^{2+} and Zn^{2+} ions, causing a change of redox current intensity measured by Sweep wave voltammetry (SWV). The change in current intensity was directly proportional to the SE's concentration in the food products. LOD of the developed immunosensor was 0.01 ng/mL and 0.02 ng/mL respectively, for Pbs-QDs and ZnS-QDs for SEB (Sharma A, et al., 2014, Sharma A, et al., 2015). Later in 2016, the same group exploited electrical properties of Graphene (Gr), GNPs, (Gr-Ch-GNPs) and multiwalled CNTs (MWCNTs) (MWCNTs-Ch-GNPs) as signal amplifier and ferrocene carboxylic acid as label in the detection of SEB. Both composite nanostructures (Gr-Ch-GNPs and MWCNTs-Ch-GNPs) provided the large surface area to immobilize more capturing MAbs. In the presence of SEB, a sandwich immune complex was formed (electrode surface-enterotoxin ferrocene labeled capturing Abs) and ferrocene moieties were oxidized to generate ferrocenium ions. The generated ions were transferred to nanostructure-modified GCE to cause an increase in redox current which got amplified by Gr-Ch-GNPs and MWCNTs-Ch-GNPs nanocomposite. The limit detection for this electrochemical immunosensor was 5 ng/mL and 10 ng/mL respectively, for Gr-Ch-GNPs and Gr-Ch-GNPs (Sharma A, et al., 2016). Unlike the previously discussed nanobiosensor, acid treatment was not required for the release of ions and SEB was detected within 35 minutes. The amalgamation of aptamer and nanoparticles in developing electrochemical nanobiosensors for food toxins detection provided highly sensitive, rapid, and cost-effective assays. (Deng R et al., 2018, Chen X et al., 2018, Xiong X et al., 2018, Nodoushan SM et al., 2019). Xiong and group 2017, developed an electrochemical aptasensor that relied on target induced inhibition of electron transfer and subsequent changes in faradic current. For the detection of SEB, thiolated SEB specific aptamer was immobilized upon the gold electrode and the change of interfacial electron transfer resistance in the presence of SEB with electrochemical impedance spectroscopy (EIS) was monitored. In the absence of SEB, non-interrupted electron transfers with small faradic impedance feedback of (Fe(CN)6)-3/-4 (ferri-ferrocynaide) was measured with low EIS signal. While in the presence of SEB, target-ligand complex formation inhibited electron transfer leading to increase in faradic impedance. EIS signal intensity increased with an increase in SEB concentration. The lowest limit of detection of the assay was 0.17 ng/mL with a linear detection range of 0.5 ng/mL to 500 ng/mL for SEB (Xiong X et al., 2017). An electrochemical aptasensor was developed by Deng R et al., 2014 by taking advantage of HCR based nucleic acid amplification technology Aptamer was liberated from DNA duplex followed by hybridization with biotinylated detecting probe and capturing probe in the presence of SEB. Biotinylated probe was detected using SA-HRP which catalyzed the peroxidase substrate to produce end point signal which was proportional to SEB concentration in the system. The LOD of assay was 0.24 ng/mL with a linear range was from 2- 512 ng/mL. Similarly, Xiong et al., 2019 explored HCR approach in the detection of SE and developed an electrochemical immunosensor with Hexaammineruthenium (III) chloride $(Ru(NH3)6)^{3+}$ as signal amplifier. Multiple nucleic acid aptamers like capturing aptamer (cpDNA) (immobilized on gold electrode), trigger DNA apatmer (form duplex with target specific aptamer), auxiliary DNA apatmer (hybridize with trigger DNA apatmer), and DNA probes form a flower-like structure, because of the target induced competition. The trigger DNA hybridized with cpDNA in the presence of the target and triggered the HCR process with auxiliary DNA molecules. The hybridized DNA molecule allowed to adsorbed countless (Ru(NH3)6)3+ molecules leading to signal amplification. The immunsosensor had a linear range from 5100 pg/mL with a LOD of 3 pg/mL for SEB (Xiong X et al., 2019).

Chen X et al., 2018 also designed an electrochemical sensor with the lowest limit of detection of 0.17 ng/mL by employing DNA pyramid frustum nanostructure (TPFDNA) (Chen X et al., 2018). Nodoushan SM and his coworkers developed an apatmer based electrochemical sensor for SEB by using rGO and gold nanourchins (AuNUs) as electrode modifiers. In this electrochemical assay hematoxylin was implied as a signal generator to obtain a LOD of 0.21 fM and the linear range of detection from 5.0 to 500.0 fM (Nodoushan SM et al., 2019).

2.3. Multiplex Sensing

Staphylococcal food poisoning can be caused by the intoxication of more than one SEs in food products. Although SEA and SEB are most common SFP causing enterotoxins, but other enterotoxins also cause SFP. Therefore, detection method for single SE cannot be sufficient to diagnose the SFP in food matrices. To accomplish this purpose, Goldman et al. explored the fluorescence of QDs as an alternative for conventional dyes for developing multiplex detection assays. The first multiplexed Quantum dot fluorescent-based ELISA (QLISA) was designed by Goldman et al., (2004) for 4 different bacterial toxins (ricin, shiga-like toxin 1, cholera toxin, and staphylococcal enterotoxin B). These toxins were concurrently detected by CdSe/ZnS core-shell dihydrolipoic acid 2 (DHLA) capped QDs bioconjugates (antitoxin antibodies). The developed sandwich immunoassay detected the fluorescence signal released from QDs at 510, 555, 590, and 610 nm for cholera toxin, ricin, shiga-like toxin 1, and SEB respectively (Goldman ER et al., 2004). Rubina AY and her coworkers (2010) developed hydrogel biochip for the detection of 7 SEs (Staphylococcal enterotoxins A, B, C1, D, E, G, and I) simultaneously in food samples. Like microarray, hydrogel biochips are contained different immobilized probes (proteins, oligosaccharides, DNA, or RNA) in an array onto glass slides. Biochips were prepared by dispersing a mixture of SEs antibodies immobilize onto microslide with hydrogel, this biochip incubated with a target analyte (antigen/antigen containing food sample). Afterwards sandwich immunocomplex formed with the addition of the biotinylated MAbs in an array and followed by interaction with fluorescently streptavidin. This led to the generation of fluorescent signal which was measured through Biochip analyzer with a laser source. This assay was reported with a detection range of 0.1-0.5 ng/mL (Rubina AY et al., 2010). More studies are required to optimize reagents and assay conditions to generate robust and reliable multiplex assay. A multiplex antibody based immunochromatographic (ICT) strip was developed by Wang W et al., 2016 for the measuring 5 SFP enterotoxins (type A, B, C, D, and E) by using GNPs conjugated MAbs.

Sandwich type immune interaction developed on ICT between respective capture antibodies-SE-GNPs conjugated MAbs, and this immune interaction results in visible signal generation. This multiplexed LFIA strip detects, 2.5 ng/mL of Staphylococcal enterotoxin type A, type B, and type C, 1 ng/mL of type D, and 5 ng/mL type E simultaneously within 15 minutes without need of any sophisticated apparatus (Wang WB et al., 2016).

To date, a smaller number of multiplex assays are designed and developed for SEs detection, although SFP cases revealed that the condition of food poisoning is caused by intoxication with more than one SEs. Hence, there is a need to develop more multiplex assays for SEs detection. Multiplex assay can be designed by employing a recognizing molecule that can recognize more than one target (class-specific molecules). Alternatively, multiple biorecognition elements like antibodies or aptamers against individual analytes can be immobilized on a solid surface to target the analytes. The hybrid use of aptamer and antibody in a single assay can be employed. Another way could be to employ various nanoparticles or fluorescent labels to allow the detection of analytes in a single platform. Integration with new technologies like portable reading devices or careful and novel designs of microfluidic chips and lateral flow assay strips can help develop point-of-care multiplex assays. Because of their miniaturization, ease of operation, and low cost, microfluidic devices and LFIAs are a better platform for implementing multiplexing of recognition entity/entities to detect multiple SEs simultaneously in the same food samples.

Conclusion

This chapter precisely focused on the future detection techniques for the Staphylococcal enterotoxin. It also summarizes the role of nanoparticles and aptamers in developing novel techniques with high sensitivity and specificity. Results are very promising and convincing, hence there is need to take these methods from lab to field. Most of reported assays are designed for targeting mostly SEB and SEA, whereas scarce reports on multiplex detection of toxins are available. Hence, it is needed to develop detection methods for other SEs like SEC, SED, SEE and SEI.

References

[1] Ali ME, Hashim U, Mustafa S, Che Man YB, and Islam KN (2012). "Gold nanoparticle sensor for the visual detection of pork adulteration in meatball formulation". *J Nanomater* 2012:103607. https://doi.org/10.1155/2012/103607

[2] Ali ME, Hashim U, Mustafa S, Che Man YB, Adam T, and Humayun Q (2011). "Nanobiosensor for the detection and quantification of pork adulteration in meatball formulation". *J Exp Nanosci* 2011, 9:152e60 (a). https://doi.org/10.1080/17458080.2011.640946

[3] Ali ME, Hashim U, Mustafa S, Che Man YB, Yusop MHM, Kashif M, Dhahi TS, Bari MF, Hakim MA, and Latif MA (2011). Nanobiosensor for detection and quantification of DNA sequences in degraded mixed meats. *J Nanomater* 2011:781098 (b). https://doi.org/10.1155/2011/781098

[4] Ali ME, Hashim U, Mustafa S, Man YB, Yusop MH, Bari MF, Islam KhN, and Hasan MF (2011). Nanoparticle sensor for label free detection of swine DNA in mixed biological samples. *Nanotechnology* 2011, 22:195503 (c). https://doi.org/10.1088/0957-4484/22/19/195503

[5] Argudín MA, Mendoza MC, and Rodicio MR (2010). "Food poisoning and *Staphylococcus aureus* Enterotoxins". *Toxins* (2010) 2:1751–1773. https://doi.org/10.3390/toxins2071751

[6] Blind M and Blank M (2015). Aptamer Selection Technology and Recent Advances. *Mol Ther. Nucleic Acids*, 2015, 4, e223. https://doi.org/10.1038/mtna.2014.74

[7] Braiek M, Rokbani KB, Chrouda A, Mrabet B, Bakhrouf A, Maaref A, and Jaffrezic-Renault N (2012). "An Electrochemical Immunosensor for Detection of *Staphylococcus aureus* Bacteria Based on Immobilization of Antibodies on Self-Assembled Monolayers-Functionalized Gold Electrode". *Biosensors*, 2012, 2: 417–426. https://doi.org/10.3390/bios2040417

[8] Campbell GA, Medina MB, and Mutharasan R (2007). "Detection of Staphylococcus enterotoxin B at picogram levels using piezoelectric-excited millimeter sized cantilever sensors". *Sens Actuators B: Chem* 2007, *126*(2), 354-360. https://doi.org/10.1016/j.snb.2007.03.021

[9] Carfora V, Caprioli A, Marri N, Sagrafoli D, Boselli C, Giacinti G, Giangolini G, Sorbara L, Dottarelli S, Battisti A, Amatiste S (2015). "Enterotoxin genes, enterotoxin production, and methicillin resistance in *Staphylococcus aureus* isolated from milk and dairy products in Central Italy". *Int Dairy J* 2015, 42, 12–15. https://doi.org/10.1016/j.idairyj.2014.10.009

[10] Chang L, Li JM, and Wang LN (2016). "Immuno-PCR: An ultrasensitive immunoassay for biomolecular detection". *Anal Chim Acta* (2016) 910:12-24. https://doi.org/10.1016/j.aca.2015.12.039

[11] Chatrathi MP, Wang J, and Collins GE (2007). "Sandwich electrochemical immunoassay for the detection of Staphylococcal Enterotoxin B based on immobilized thiolated antibodies". *Biosens Bioelect* 2007, *22*(12):2932-2938. https://doi.org/10.1016/j.bios.2006.12.03

[12] Chen X, Shi X, Liu Y, Lu L, Lu Y, Xiong X, Liu Y, and Xiong X (2018). "Impedimetric determination of Staphylococcal enterotoxin B using electrochemical switching with DNA triangular pyramid frustum nanostructure". *Microchimica Acta* (2018) 185:460. https://doi.org/10.1007/s00604-018-2983-3

[13] Deng R, Li Wang L, Yi G, Hua E, and Xie G (2014). "Target-induced aptamer release strategy based on electrochemical detection of staphylococcal enterotoxin B using GNPs-ZrO2-Chits film". *Colloids Surf B* (2014). http://dx.doi.org/ 10.1016/j.colsurfb.2014.04.02

[14] Ding T, Yu YY, Schaffner DW, Chen SG, Ye XQ, Liu DH (2015). "Farm to consumption risk assessment for *Staphylococcus aureus* and staphylococcal enterotoxins in fluid milk in China". *Food Control* (2016) 59:636-64. http:// dx.doi.org/10.1016/j.foodcont.2015.06.049

[15] Dirks RM and Pierce NA (2004). "Triggered amplification by hybridization chain reaction". *Proc Nat. Acad Sci U S A*, 2004, 101, 15275–15278. https://doi.org/ 10.1073/pnas.0407024101

[16] Evenson ML, Hinds MW, Bernstein RS, Bergdoll MS, (1988). "Estimation of human dose of Staphylococcal enterotoxin A from a large outbreak of staphylococcal food poisoning involving chocolate milk". *Int J Food Microbiol* 7 (1988) 311–316. https://doi.org/10.1016/01681605(88)90057-8

[17] Farka Z, Jurik T, Kovar D, Trnkova L, Skladal P (2017). "Nanoparticle based immunochemical biosensors and assays: Recent advances and challenges". *Chem Rev 2017*, 117, 9973–10042. https://doi.org/10.1021/acs.chemrev.7b00037

[18] Fusco V., Quero, G.M., Morea, M., Blaiotta, G., Visconti, A (2011). "Rapid and reliable identification of *Staphylococcus aureus* harbouring the enterotoxin gene cluster (egc) and quantitative detection in raw milk by real time PCR". *Int J Food Microbiol* (2011) 144:528-537. https://doi.org/10.1016/j.ijfoodmicro.2010.11.016

[19] Goldman ER, Clapp AR, Anderson GP, Uyeda HT, Mauro JM, Medintz IL, Mattoussi, H (2004). "Multiplexed toxin analysis using four colors of quantum dot fluororeagents". *Anal Chem* 2004, 76:684–688. https://doi.org/10.1021/ac035083r

[20] Guo Y, Wei B, Xiao S, Yao D, Li H, Xu H, Song T, Li X, and Liang H (2017). "Recent advances in molecular machines based on toehold mediated strand displacement reaction". *Quantitative Biology* 2017, 5 (1): 25–41. doi:10.1007/ s40484-017-0097-2.

[21] Guo Y, Zhang Y, Pei R, Cheng Y, Xie Y, Yu H, Yao W, Li HW, and Qian H (2019). "Detecting the adulteration of antihypertensive health food using G-insertion enhanced fluorescent DNA-AgNCs". *Sens Actuators B Chem* 2019, 281:493–498. https://doi.org/10.1016/j.snb.2018.10.101

[22] Haddada MB, Hu D, Salmain M, Zhang L, Peng C, Wang Y, Liedberg B, Boujday S (2017). "Gold nanoparticle-based localized surface plasmon immunosensor for staphylococcal enterotoxin A (SEA) detection". *Anal Bioanal Chem* 2017. https://doi.org/10.1007/s00216017-0563-8

[23] Heidinger JC and Winter CK (2009). "Quantitative microbial risk assessment for *Staphylococcus aureus* and Staphylococcus enterotoxin A in raw milk". *J Food Prot* (2009) 72:1641–1653. https://doi.org/10.4315/0362-028x-72.8.1641

[24] Hennekinne MJ, De S, and Buyser D (2012). "*Staphylococcus aureus* and its food poisoning toxins: characterization and outbreak investigation". *FEMS Microbiol Rev* (2012) 36:815–836. https://doi.org/10.1111/j.1574-6976.2011.00311.x

[25] Kaittanis C, Santra S, Perez JM (2010). "Emerging nanotechnology based strategies for the identification of microbial pathogenesis". *Adv Drug Deliv Rev* 2010, 62:408e23. https://doi.org/10.1016/j.addr.2009.11.013

[26] Kasai S, Yokota A, Zhou H, Nishizawa M, Niwa K, Onouchi T, and Matsue T (2000). "Immunoassay of the MRSA-related toxic protein, leukocidin, with scanning electrochemical microscopy". *Anal Chem* 2000, 72, 5761–5765. https://doi.org/10.1021/ac000895y

[27] Kim MS, Kweon SH, Cho S, An SSA, Kim MI, Doh J, and Lee J (2017). Pt-Decorated Magnetic Nanozymes for Facile and Sensitive Point-of-Care Bioassay. *ACS Appl Mater Interfaces* 2017, 9:35133–35140. https://doi.org/10.1021/acsami.7b12326

[28] Kotb M, and Fraser JD editors (2008). "Superantigens: molecular basis for their role in human diseases". Washington: ASM Press. *Emerg Infect Dis* 2008 14(5): 866–867. https://doi.org/10.3201/eid1405.080089

[29] Larkin, E.A., Carman, R.J., Krakauer, T., Stiles, B.G. *Staphylococcus aureus* the toxic presence of a pathogen extraordinaire *Curr Med Chem* 16 (2009), pp. 4003-4019. https://doi.org/10.2174/092986709789352321

[30] Leonard P, Hearty S, Brennan J, Dunne L, Quinn J, Chakraborty T, O'Kennedy R (2003). "Advances in biosensors for detection of pathogens in food and water". *Enzyme Microb Technol* 2003 32:3e13. https://doi.org/10.1016/S0141-0229(02)00232-6

[31] Lin HC and Tsai WC (2003). "Piezoelectric crystal immunosensor for the detection of Staphylococcal enterotoxin B". *Biosens Bioelect* 2003 *18*(12):1479-1483. https://doi.org/10.1016/S0956-5663(03)00128-3

[32] Liu M, Xu J, Yang F, Gu Y, Chen H, Wang Y, and Li F (2017). "Sensitive electrochemical detection of DNA damage based on in situ double strand growth via hybridization chain reaction". *Anal Bioanal Chem* 2017, 409, 6821–6829. https://doi.org/10.1007/s00216-0170641-y

[33] Liu P, Yang X, Sun S, Wang Q, Wang K, Huang J, Liu J, and He L (2013). "Enzyme-Free Colorimetric Detection of DNA by Using Gold Nanoparticles and Hybridization Chain Reaction Amplification". *Anal Chem* 2013, 85:7689–7695. https://doi.org/10.1021/ac4001157

[34] Lopez BP and Merkoci A. "Nanomaterials based biosensors for food analysis applications". *Trends Food Sci Technol* 2011, 2:625e39. https://doi.org/10.1016/j.tifs.2011.04.001

[35] Lu SS, Hu T, Wang S, Sun J, and Yang X (2016). "Ultra-Sensitive Colorimetric Assay System Based on the Hybridization Chain Reaction Triggered Enzyme Cascade Amplification". *ACS Appl Mater Interfaces*, 2016, 9:167–175. https://doi.org/10.1021/acsami.6b13201

[36] Marrack P and Kappler J (1990). "The staphylococcal enterotoxins and their relatives". *Science* 1990, 248:705–711. https://doi.org/10.1126/science.2185544.

[37] Mondal B, Ramlal S, Lavu PS, N B and Kingston J (2018). "Highly Sensitive Colorimetric Biosensor for Staphylococcal Enterotoxin B by a Label-Free Aptamer and Gold Nanoparticles". *Front Microbiol* 2018 9:179. https://doi.org/10.3389/fmicb.2018.00179

[38] Mondal B, Ramlal S, Rani Lavu PS, Murali HS, Batra HV (2015). "A combinatorial systematic evolution of ligands by exponential enrichment method for selection of aptamer against protein targets." *Appl Microbiol Biotechnol* 2015, 99(22):9791-9803. https://doi.org/10.1007/s00253015-6858-9

[39] Nagaraj S, Ramlal S, Kingston J, Batra HV (2016). "Development of IgY based sandwich ELISA for the detection of staphylococcal enterotoxin G (SEG), an egc toxin: *Int J Food Microbiol* (2016) 237 136-141. https://doi.org/10.1016/j.ijfoodmicro.2016.08.009

[40] Nodoushan SM, Nasirizadehb N, Amania J, Halabiana R, Imani Fooladi AA (2019). "An electrochemical aptasens or for staphylococcal enterotoxin B detection based on reduced graphene oxide and gold nanourchins". *Biosens Bioelectron* (2019) 127:221–228. https://doi.org/10.1016/j.bios.2018.12.021

[41] Nunes MM and Caldas ED (2017). "Preliminary quantitative microbial risk assessment for staphylococcus enterotoxins in fresh minas cheese, a popular food in Brazil". *Food Control* 2017, 73:524–531. https://doi.org/10.1016/j.foodcont.2016.08.046

[42] Odom TW, Huang JL, Kim P, Lieber CM (1998). "Atomic structure and electronic properties of single-walled carbon nanotubes". *Nature* 1998, **391**:62–4. https://doi.org/10.1038_34145

[43] Pinheiro L, Brito CI, de Oliveira A, Martins PYF, Pereira VC, da Cunha M (2015). "*Staphylococcus epidermidis* and *Staphylococcus haemolyticus*: Molecular Detection of Cytotoxin and Enterotoxin Genes". *Toxins* (2015) 7:3688-3699. https://doi.org/10.3390/toxins7093688

[44] Rubina AY, Filippova MA, Feizkhanova GU, Shepeliakovskaya AO, Sidina EI, Boziev KM, Laman AG, Brovko FA, Vertiev Yu. V, Zasedatelev AS, and Grishin, EV (2010). "Simultaneous Detection of Seven Staphylococcal Enterotoxins: Development of Hydrogel Biochips for Analytical and Practical Application". *Anal Chem* 2010 82(21):8881–8889. https://doi.org/10.1021/ac1016634

[45] Scallan, E., Hoekstra, R.M., Angulo, F.J., Tauxe, R.V., Widdowson, M.A., Roy, S.L., Jones, J.L., Griffin, P.M. Foodborne Illness Acquired in the United States-Major Pathogens, *Emerging Infect Dis* (2011) 17:7-15. https://doi.org/10.3201/eid1701.P11101

[46] Seo JH, Park BH, Oh SJ, Choi G, Kim DH, Lee, EY, Seo, TS (2017). "Development of a high-throughput centrifugal loop-mediated isothermal amplification microdevice for multiplex foodborne pathogenic bacteria detection". *Sensor Actuat B* (2017) 246:146-153. http://dx.doi.org/10.1016/j.snb.2017.02.051

[47] Sharma A, Rao VK, Kamboj DV, Gaur R, Shaik M and Srivastava A (2016). "Enzyme free detection of Staphylococcal Enterotoxin B (SEB) using ferrocene carboxylic acid labeled monoclonal antibodies: An electrochemical approach". *New J Chem* 2016. http://dx.doi.org/10.1039/C5NJ03460D

[48] Sharma A, Rao VK, Kamboj DV, Gaur R, Upadhyay S, Shaik M (2015). "Relative efficiency of zinc sulfide (ZnS) quantum dots (QDs) based electrochemical and fluorescence immunoassay for the detection of Staphylococcal enterotoxin B (SEB)". *Biotechnol Rep* (2015). http://dx.doi.org/10.1016/j.btre.2015.02.004

[49] Sharma A, Rao VK, Kamboj DV, Upadhyay S, Shaik M, Shrivastava AR, and Jain R (2014). "Sensitive detection of staphylococcal enterotoxin B (SEB) using quantum dots by various methods with special emphasis on an electrochemical immunoassay approach". *RSC Adv* 2014, 4, 34089. https://doi.org/10.1039/C4RA02563F

[50] Shyu R-H, Tang S-S, Chiao D-J, and Hung Y-W (2010). Gold nanoparticle-based lateral flow assay for detection of staphylococcal enterotoxin B. *Food Chemistry* (2010) 118:462–466. https://doi.org/10.1016/j.foodchem.2009.04.106

[51] Sonawane SK, Arya SS, LeBlanc JG, and Jha N (2014). Use of nanomaterials in the detection of food contaminants. *Eur J Nutr Food Saf* 2014, 4:301e17. https://doi.org/10.1016/B978-0-12-811942-6.00015-7

[52] Sospedra I, Marın R, Manes J, and Soriano JM (2012). "Analysis of staphylococcal enterotoxin A in milk by matrix-assisted laser desorption/ionization-time of flight mass spectrometry". *Food Chem* 2012, 133:163–166 (a). https://doi.org/10.1007/s00216-011-4906-6

[53] Sospedra I, Soler C, Mañes J, and Soriano JM (2012). "Rapid whole protein quantitation of staphylococcal enterotoxins A and B by liquid chromatography/mass spectrometry". *J Chromatogr A* 2012, 1238, 54–59 (b). https://doi.org/10.1016/j.chroma.2012.03.022

[54] Stoltenburg R, Reinemann C, and Strehlitz B (2007). "SELEX--a (r)evolutionary method to generate high-affinity nucleic acid ligands". *Biomol Eng* 2007 24:381–403. https://doi.org/10.1016/j.bioeng.2007.06.001

[55] Tallent SM, DeGrasse JA, Wang NY, Mattis DM, and Kranz, DM (2013). Novel Platform for the Detection of *Staphylococcus aureus* Enterotoxin B in Foods. *Appl Environ Microbiol* 79 (2013) 1422-1427. https://doi.org/10.1128/AEM.02743-12

[56] Tan F, Xie X, Xu A, Deng K, Zeng Y, Yang X, and Huang H (2018). "Fabricating and Regulating peroxidase-like activity of eggshell membrane-templated gold nanoclusters for colorimetric detection of staphylococcal enterotoxin B". *Talanta* 2018. https://doi.org/10.1016/j.talanta.2018.10.031

[57] Tang D, Tang J, Su B, and Chen G (2010). "Ultrasensitive Electrochemical Immunoassay of Staphylococcal Enterotoxin B in Food Using Enzyme-Nanosilica-Doped Carbon Nanotubes for Signal Amplification". *J Agric Food Chem* 2010, 58:10824–10830. https://doi.org/10.1021/jf102326m

[58] Tatini SR (1976). Thermal stability of enterotoxins in food. *J Milk Food Technol* 1976 39: 432-438. https://doi.org/10.4315/0022-2747-39.6.432

[59] Thipmanee O, Numnuam A, Limbut W, Buranachai C, Kanatharana P, Vilaivan T, Hirankarn N and Thavarungkul P. "Enhancing capacitive DNA biosensor performance by target overhang with application on screening test of HLA-B*58:01 and HLA-B*57:01 genes". *Biosens Bioelectron* 2016, 82:99–104. https://doi.org/10.1016/j.bios.2016.03.065

[60] Tran Q, Nguyen HTHH, Mai An T, Nguyen TT, Khue Vu Q, and Phan TN (2012). "Development of electrochemical immunosensors based on different serum antibody immobilization methods for detection of Japanese encephalitis virus: *Adv Na. Sci: Nanosci Nanotechnol* 2012, 3:1–6. https://doi.org/10.1088/2043-6262/3/1/015012

[61] Upadhyay N and Nara S (2018). "Lateral flow assay for rapid detection of *Staphylococcus aureus* enterotoxin A in milk". *Microchem J* (2018) 137:435–442. https://doi.org/10.1016/j.microc.2017.12.011

[62] Valdes MG, Gonz alez ACV, Calz on JAG, Dı az-Garcıa ME (2009). "Analytical nanotechnology for food analysis". *Microchim Acta* 2009,166:1e19. https://doi.org/10.1007/s00604-009-0165-z

[63] Wang H, Wang DM, Gao MX, Jian Wang J, and Huang CZ (2014). "Potassium-induced G-quadruplex DNAzyme as a chemiluminescent sensing platform for highly selective detection of K^{+}". *Anal Methods* 2014 6,:7415-7419. https://doi.org/10.1039/C4AY01411A

[64] Wang WB, Liu LQ, Xu LG, Kuang H, Zhu JP, and Xu CL (2016). "Gold-Nanoparticle-Based Multiplexed Immunochromatographic Strip for Simultaneous Detection of Staphylococcal Enterotoxin A, B, C, D, and E". *Part Part Syst Charact* (2016) 33: 388-395. https://doi.org/10.1002/ppsc.201500219

[65] Wu K-H, Huang W-C, Shyu R-H, and Chang S-C (2020). "Silver nanoparticle-base lateral flow immunoassay for rapid detection of Staphylococcal enterotoxin B in milk and honey". *J Inorg Biochem* 2020, 210:111163. https://doi.org/10.1016/j.jinorgbio.2020.111163

[66] Wunderlichova L, Bunkova L, Koutny M, Jancova P, Bunka F (2014). "Formation, Degradation, and Detoxification of Putrescine by Foodborne Bacteria: A Review". *Compr Rev Food Sci F* (2014), 13:1012-1030. https://doi.org/10.1111/1541-4337.12099

[67] Xie X, Tan F, Xu A, Deng K, Zeng Y, and Huang H (2018). "UV induced peroxidase-like activity of gold nanoclusters for differentiating pathogenic bacteria and detection of enterotoxin with colorimetric readout". *Sens Actuators: B. Chem* (2018). https://doi.org/10.1016/j.snb.2018.10.019

[68] Xiong X, Luo Y, Lu Y, Xiong X, Li Y, Liu Y, and Lu l (2019). "Ultrasensitive detection of Staphylococcal enterotoxin B in milk based on target-triggered assembly of the flower like nucleic acid nanostructure". *RSC Adv* 2019, 9:42423–42429. https://doi.org/10.1039/c9ra08869e

[69] Xiong X, Shi X, Liu Y, Lu L, and You J (2017). "Chiroplasmonic assemblies of gold nanoparticles as a novel method for sensitive detection of alpha-fetoprotein". *Anal Methods*, 2017. https://doi.org/10.1039/C7AY02452E

[70] Xu Y, Huo B, Sun X, Ning B, Peng Y, Bai J, and Gao Z (2018). "Rapid detection of staphylococcal enterotoxin B in milk samples based on fluorescence hybridization chain reaction amplification". *RSC Adv* 2018, 8:16024–16031. https://doi.org/10.1039/C8RA01599F

[71] Xuan F, Fan TW, and Hsing I-M (2015). "Electrochemical Interrogation of Kinetically-Controlled Dendritic DNA/PNA Assembly for Immobilization-Free

and Enzyme-Free Nucleic Acids Sensing". *ACS Nano* 2015 9: 5027–5033. http://dx.doi.org/10.1021/nn507282f

[72] Yang M, Kostov Y, Rasooly A (2008). "Carbon nanotubes based optical immunodetection of staphylococcal enterotoxin B (SEB) in food". *Int J Food Microbiol* 2008,127:78e83. https://doi.org/10.1016/j.ijfoodmicro.2008.06.012

[73] Yang M, Kostov Y, Bruck HA, and Rasooly A (2009). "Gold nanoparticle-based enhanced chemiluminescence immunosensor for detection of Staphylococcal Enterotoxin B (SEB) in food". *Int J Food Microbiol* 2009, *133*(3):265-271. 10.1016/j.ijfoodmicro.2009.05.029

[74] Yurke B and Millis AP (2003). "Using DNA to power nanostructures". *Genet Program Evolvable Machs* 2003, 4 (2): 111–122. http://dx.doi.org/10.1023/A:1023928811651

[75] Yang MH, Javadi A, Li H, Gong SQ (2010). "Ultrasensitive immunosensor for the detection of cancer biomarker based on graphene sheet". *Biosens Bioelect* 2010, 26:560–5. http://dx.doi.org/10.1016/j.bios.2010.07.040

[76] Zhang DY and Seelig G (2011). "Dynamic DNA nanotechnology using strand-displacement reactions". *Nature Chemistry* 2011, 3(2): 103–13. http://dx.doi.org/10.1038/nchem.957.

[77] Zhang X, Khan IM, Ji H, Wang Z, Tian H, Cao W, Mi W (2020). "A Label-Free Fluorescent Aptasensor for Detection of Staphylococcal Enterotoxin A Based on Aptamer-Functionalized Silver Nanoclusters". *Polymers* 2020 12:152. http://dx.doi.org/10.3390/polym12010152

[78] Zhao J, Chen C, Zhang L, Jiang J, and Yu R. "An electrochemical aptasensor based on hybridization chain reaction with enzyme-signal amplification for interferon-gamma detection". *Biosens Bioelectron* 2012, 36 (1):129–134. https://doi.org/10.1016/j.bios.2012.04.013

[79] Zhao Y, Zhang Q, Meng Q, Wu F, Zhang L, Tang Y, and Guan Y, and An L (2017). "Quantum dots-based lateral flow immunoassay combined with image analysis for semiquantitative detection of IgE antibody to mite". *An Int J Nano-medicine* 2017 12:4805-4812. https://doi.org/10.2147/IJN.S134539

[80] Zhou D, Xie G, Cao X, Chen X, Zhang X, Chen H (2016). "Colorimetric determination of staphylococcal enterotoxin B via DNAzyme-guided growth of gold nanoparticles". *Microchim Acta* 2016 183:2753–2760. https://doi.org/10.1007/s00604-016-1919-z

[81] Zhu J, Zhang L, Teng Y, Lou B, Jia X, Gu X, and Wang E (2015). "G-quadruplex enhanced fluorescence of DNA–silver nanoclusters and their application in bioimaging". *Nanoscale* 2015 7:13224–13229. https://doi.org/10.1039/C5NR03092G

Chapter 4

The Implications of Nanobiosensors in Agriculture for Human Welfare

Trupti K. Vyas[1,*]
Mansi Mehta[2]
and Nilima Karmkar[3]
[1]Food Quality Testing Laboratory, N M College of Agriculture,
Navsari Agricultural University, Gujarat, India
[2]Department of Biotechnology, Veer Narmad South Gujarat university, Gujarat, India
[3]Department of Biochemistry, NMCA, Navsari Agricultural University, Gujarat, India

Abstract

Farmers and the agricultural industry have faced significant difficulties in past few decades as a result of climate change. Biotic and abiotic stresses have emerged, such as salinity, rising global temperatures, plant pathogen-related stress, and water scarcity, which negatively affect plant health. It causes severe damage to crops, resulting in economic losses to farmers. Thus, there is a great need for technologies that can detect biotic and abiotic stresses on the farm. One such promising advancement in novel technology is nanobiosensor. A biological recognition molecule is typically attached on the surface of a signal transducer in nanobiosensors. Because the response between the biorecognition molecule and the analyte is heterogeneous, the design of the biosensing interface is critical in defining the nanobiosensor's performance. They can analyse both quantitative and qualitative bio elements through the formation of observable signals as they are analytical equipment. Nanobiosensors are very selective, precise and provide results in very short time with great

* Corresponding Author's Email: vyastrupti@hotmail.com.

In: Biosensing
Editor: Rushika Patel
ISBN: 979-8-88697-911-4

linearity and high sensitivity. Thus, in agriculture it can be used to detect plant pathogen interactions, early disease diagnosis, abiotic stress detection and quality control. The present chapter provides novel nanobiosensor and its application in agriculture.

Keywords: nanobiosensor, nano particles, agriculture, biofertilizer, indicator

1. Introduction

India's economy heavily relies on agriculture. However, with a growing population and limited agricultural land, farmers face significant challenges in meeting the rising demand for food (Das and Das 2019; Rockström et al., 2017). The trend of urbanization has resulted in the encroachment of agricultural land, exacerbating the scarcity of resources for farming. Moreover, increasing soil salinity is another issue in limiting the agricultural farming land (Ghassemi et al., 1995; Alexandratos, 2009; Dagar et al., 2016). Over the past few decades, climate change has caused unpredictable rainfall patterns, flooding, and other environmental factors that increase stress on plants. Additionally, water scarcity has become a concern, highlighting the need for sustainable approaches to predict and manage plant stress, disease, and other related issues in the agricultural field.

As a result of global warming, temperatures are on the rise, causing increased stress on plants and ecological systems. Even a small increase of 1°C can lead to a significant reduction in crop productivity. (Iizumi et al., 2017; Zhao et al., 2017). Paradigm shifts in the study of weather patterns beyond the common threshold would theoretically cause a sudden drop in temperature in tropical and subtropical zones, potentially resulting in a decrease in global food production (Budhathoki and Zander 2019; Thakur and Nayyar 2013).

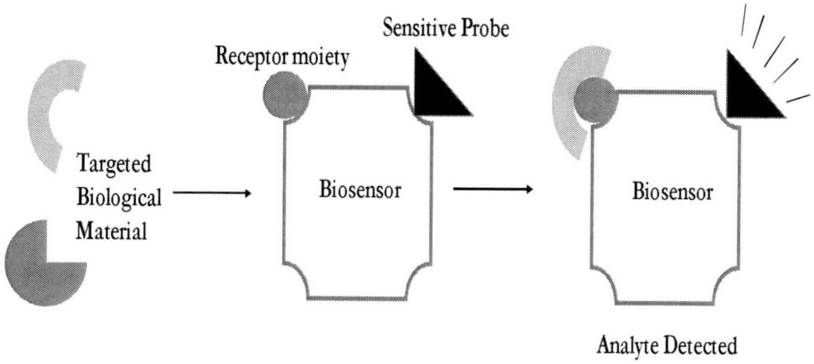

Figure 1. A schematic representation of biosensor-based activity.

Farmers and agricultural personnel require tools that allow them to anticipate and identify stress or disease in the field, enabling them to promptly eliminate the pathogen or causative agent and prevent on-site losses. These can be only possible when farmers able to identified the early stage of infection/disease. But all these required the collection of sample followed by laboratory analysis, which is time consuming and delayed the remedial application in the field. Thus, there is great urge for the tools which can be easily operated by the farmers on the field. Biosensors which are smaller devise and can detect targeted molecules in short period of time, which can overcome above problems to save economic loss.

Recently, nano biotechnology gave a hope in the field of science and technology. Due to advancement in nano science, scientists develop some tools to predict/detect the molecule(s) which are helpful in prediction and forecast. One such devise is nano biosensors. Figure 1 describes basic biosensor activity to analyse biological materials and detect result based on its sensitive and confirmational change. Biosensors are the tool which comprised a biological substance like enzymes, antibodies, receptors, tissues or DNA of microbial origin molecules to identify or quantify the analytes. Analytes can be measured by electrical, chemical or physical signal generated using nano biosensors (Walia et al., 2018).

2. Types of Nano Biosensors

Advancement in the biosensors has mainly accomplished by improvement in the detection techniques. There are two types of biosensors to detect analytes are label-free and label-based techniques. In label-based techniques fluorescence, chemiluminescence or radioactive tag or label use to detect analytes (Harz et al., 2011; Li et al., 2008; Ma et al., 2014; Akshath et al., 2012; Gibson et al., 2011). While label-free techniques are a novel technique that do not required labeling or tagging of ligand another one.

Label-free detection is generally used when tagging of target molecules is difficult. However, with the advancement in biotechnological tools and techniques, development of label free biosensors becomes easy. Such several label-free biosensors are developed based on field-effect transistors-based biosensors, silicon nanowire field-effect transistor biosensors, magnetoelastic biosensors, optical-based biosensors, surface stress-based biosensors etc. Present chapter describe about the types and application of nanobiosensors in agriculture (Table 1).

Table 1. Types of Nano Biosensors and its application agriculture

Sensor Type	Nano material used	Application	References
Fluorescence	Quantum dots	Detection of pathogens	Esker et al., (2018)
	Quantum dots	Phytoplasma aurantifolia	Rad et al., (2012)
	Receptor DAD2 from Petunia hybrid and HTLT from Striga hermohthica with green fluorescent protein	Detection of strigolactones as signalling molecules for plant growth and parasitism	Chesterfield et al., 2020
Electrochemical	Single and Multiwalled carbon nanotubes	Detection of pesticides methyl parathion, parathion and paraoxan	Fu, Q. et al., (2019)
	DNA based biosensor	Detection of Phytophthorapulmivora causing black pod rot in cacao pod	Wong et al., (2017)
	Calcium phosphate	Pathogenic detection	James et al., (2019)
	Graphene based with molecular imprinted polymers	Pesticide detection of chlorothalonil and chlorpyrifos methyl	Manjunatha et al., (2016)
	Mesoporous molecular sieves embedded with carbon dots CDs@SBA15@MIP	Detection of kaempferol polyphenol in vegetables	Negrete, J. C. (2020)
	Zinc oxide and Copper	Enhance the germination of tomato chili and cucurbits in Mexico	Xu et al., (2020)

Sensor Type	Nano material used	Application	References
Immunosensor	Quantum dot nanosensor	Detection of mycotoxins ZEA, DON, FB1/FB2 in corn oats and barley	He et al., (2020)
	Carbon nanotubes	*Ganoderma boninse* for infection of palm oil tree	Negrete (2020)
Differential pulse voltammetry	Enzyme	Detection of paraoxon	Lattanzio et al., (2012)
	Cholinergic enzyme	Detection of dichlorvos	Safarpour et al., (2012)
Amperometric	Cholinergic enzyme	Detection of chlorpyrifos	Singh et al., (2010)
	Enzyme	Detection of Malathion	Mahmoudi et al., (2019)
Square wave voltammetry	Cholinergic enzyme	Detection of malathion	Umasankar et al., (2013)
Coloured reaction	Enzym0065	Detection of chlorpyrifos	Cui et al., (2018a,b),
Volatile organic compound profile	Carbon nanomaterials	Detection of pathogens depending on the organic compounds released	Mishra et al., (2021)

2.1. Label-Based Detection

It is feasible to assess analytes in traces by dissolving nanoparticle labels and measuring dissolved ions by voltammetry stripping, a potent analytical technique in determining the impacts of metals (Wang G 2014). On immunoassays, most label-based detection methods are built. Antigen antibody responses are the basis for these immunoassays. To measure the amount of protein in blood, modern immunoassays use enzyme-linked immunosorbent assay (ELISA) testing. To identify whether an antibody or antigen is present in the sample, ELISA employs a biochemical approach. For the past 30 years, ELISA has been the accepted diagnostic method. In order to improve detection, ELISA employs fluorophore-tagged linker molecules, yet these fluorophores may cause the proteins to become denatured. The presence of fluorescent markers that produce light at particular wavelengths and the augmentation or reduction of the optical signal, as in fluorescence resonance energy transfer, establish whether the binding reaction has occurred. There are numerous wash phases, which lengthens the process. ELISA is a laboratory-based approach that requires experienced, qualified professionals to conduct the tests, which causes significant delays. Due to these limitations, ELISA is inappropriate for quick point-of-care biosensing applications (Bard A. 1998).

Nanoparticle-tagged biomolecules continue to function biologically. The analyte concentration is determined the binding of the analyte receptor and the electrochemical detection of the nanoparticles (Keerthy D. 2014).

2.2. Label-Free Detection Methods

Recent developments in microfabrication and nanotechnology have aided in the development of quick point-of-care diagnostics. There are a few methods that make it possible to identify biomolecules without labels, and they are listed below. High surface energy nanoparticles are more active than bulk materials and occasionally exhibit distinct chemical characteristics.

3. Electrochemical and Electrical Detection

By automating agricultural machinery with the use of cutting-edge technology, agriculture and communication technology can contribute to the development of environmentally friendly farming practices. By utilizing sensor and information processing technology, it is possible to enhance the forecasting of crop growth and climate conditions, which can aid in the advancement of innovative fertilization techniques. Additionally, electrochemical sensor technology offers a lower sample size, greater measurement precision, and less environmental interference. In the agricultural sector, where real-time monitoring is crucial, the ease of measurement and real-time detection has a key advantage.

The most promising applications for biosensors are those requiring minimal size, convenience of use, and cost, such as point-of-care diagnostics or patient care at the bedside. Most of the time, these biosensors use changes in currents and/or voltages to measure the binding event. The method used to make the electrical measurement, such as voltametric, amperometric/coulometric, and impedance measurements, can further categories biosensors based on the use of electrochemical techniques. Electrical biosensors are extremely similar to the electroanalytical methods that have been established in the chemistry field (Bard A. et. al. 1985).

The electrochemical sensor can monitor changes in the growing condition of plants in real time, thus it is feasible to respond in a preventive process, whereas the present analytical methods assess the current results. The monitoring of plant growth data, pre and post pesticide/insecticide effects, fertilization content, moisture content, pH parameters of soil, and environmental pollution detection can all be done quickly and accurately with one portable device once this electrochemical sensor technology is commercialized in earnest. The overall electrochemical impact on the development of sustainable agriculture is shown in Figure 2.

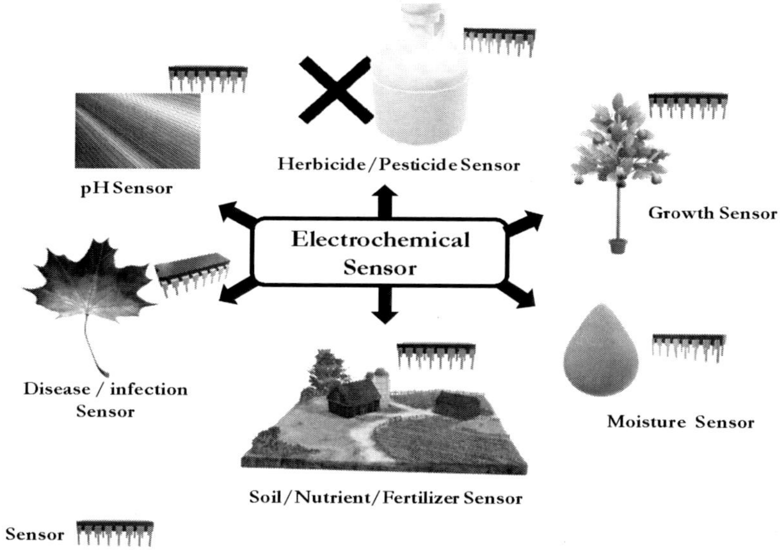

Figure 2. Electrochemical sensors for sustainable agriculture.

4. Optical Detection

The study of producing and using light to picture, detect, and work with biological materials is known as bio photonics. The confluence of photonics and biology represents a fascinating new horizon. The basic operational principle of this technology is to detect and image biological systems at molecular, cellular and organismal levels using scattering and penetrating light (Olkhov et al., 2012). In Figure 3, it is shown how a photonic sensor can use several types of molecules to produce a signal utilising scattering, refractive index, or optical absorbance. These molecules are commonly used as fertiliser, pesticides, dietary supplements, and even moisture content. A unique component of bio photonics results from the fact that the light produced by metabolic processes in living things also serves as an effective optical way of reflecting the composition and operation of living cells and organisms.

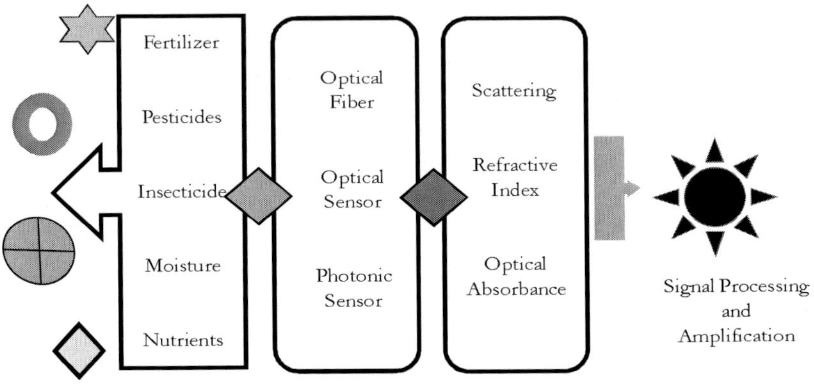

Figure 3: Schematic representation of optical biosensor.

When a chemical reaction occurs, it can emit chemiluminescence, which is the production of energy in the form of light. Such chemiluminescence occurs in synthetic compounds when a highly oxidised species, like peroxide, emits energy during a chemical reaction. The energy emitted as light when some synthetic compounds are combined with a biomolecule to create a conjugate could be employed as a detection method. A method for measuring surface activity is called SPR (Surface Plasmon Resonance). In this method, a longitudinal wave propagates along the metal surface and the dielectric contact at a specific charge density. The evanescent wave or field produced at the surface of a dense media (metal) such as gold or silver due to total internal reflection of light can pair with free electrons known as plasmons on the surface and create a resonance wave, which is recorded. As a result, an SPR adsorption profile is formed if there is a layer of biomolecules on the surface of the metal, and because of the characteristics of biomolecules, each SPR spectrum is unique. The aspect of the SPR that varies with each molecule is its angle (Bondar V. S. 2004). The SPR signal is significantly amplified when metal nanoparticles are used rather than thin films, enabling ultrasensitive detection. SPR has already been put on the market, but this method's biosensors are still in the works. Specificity and a very low signal-to-noise ratio are problems with SPR-based biodetection devices.

5. Mass-Based Detection

Cantilever sensors with micro- or nanoscale resolutions are typically used to detect mechanical properties of biological substances. To determine the mass change on the surface, the cantilever is mechanically activated to resonate at its resonant frequency. Then, the resonant frequency is compared to the frequency of the cantilever without biomolecules bound to its surface. Detecting interacting substances without labelling the biomolecules is made possible by cantilever sensors. The intricacy of both the equipment and the detection system is its main flaw. These issues stop the use of this highly accurate measurement technique for portable devices, together with the extremely high sensitivity to temperature and other factors.

6. Application of Nanobiosensor in the Agriculture

In India, most of the population depends on agriculture for their livelihood. Hence, use of advance techniques for detection of disease and stress tolerance in plant prevents major yield loss in agriculture. For sensing soil pH, moisture, a wide range of pathogens, plant hormones, plant metabolites, pesticides, herbicides, fertilisers, and metal ions, nanobiosensors can be utilised successfully in agriculture (Figure 4). Integrating nanobiosensors in a responsible and controlled manner can help sustainably increase agricultural output. Additionally, it can aid in the controlled use of agricultural inputs, hence reducing cultivation costs and pollution.

6.1. Soil Productivity

In agriculture soil fertility play a major role in plant productivity. Certain soil criteria such as pH, moisture, nutrient and pesticide residues are important parameter to determine the soil quality. Moreover, presence of plant growth promoting bacteria for nitrogen fixing ability, phosphate solubilizing ability provide information regarding good soil quality. Application of nano biosensors for detecting soil quality helps the farmers to input fertilizer and pesticides more precisely (Bellingham 2011; Prasad et al., 2017) and thus saving unnecessary application of agrochemicals and ultimately restore the ecosystem.

One of the signaling molecules that released by the plant is strigolactones. Strigolactones play a role in plant development and plant parasitism. Chesterfield and co-worker 2020 have developed a green fluorescent protein based nano biosensor using strigolactone receptors DAD2 from Petunia hybrida, and HTL7 from Striga hermonthica. This sensor used for detection of strigolactone and it is signaling pathways.

Urea is main nitrogen source for agricultural practitioner. However, urea undergoes hydrolysis and converted into ammonia which can affect seed germination and younger plants. Hence, quantification of urea in soil is important parameter for soil quality. There are several methods for detection urea, but they are pH sensitive. Deng and co-worker (2016) develops a gold nanoparticle based ultrasensitive assay for detection of urea and urease. The LOD for urea and urease is 5 μM and 1.8 U/L respectively.

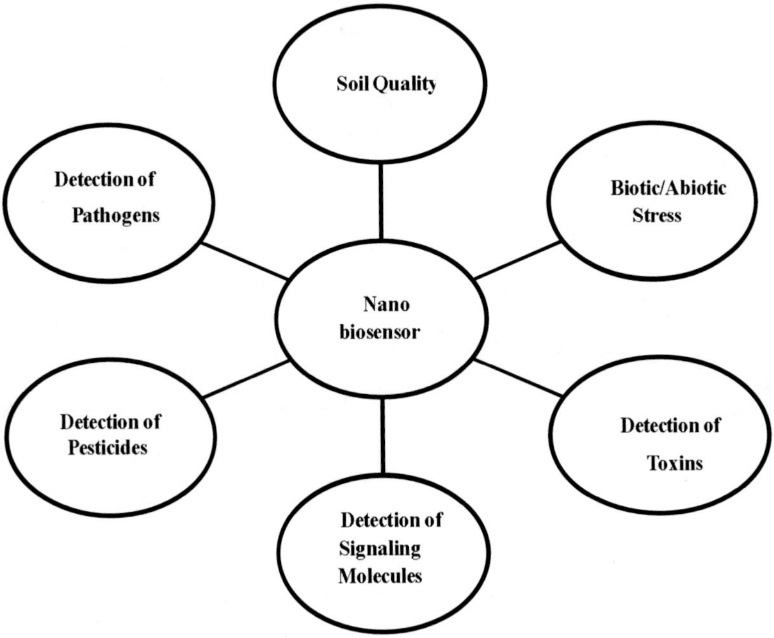

Figure 4. Application of nano biosensor in agriculture.

Monreal et al., (2015) suggested the quality of the soil can be evaluated, based on the interaction between microorganisms in the rhizosphere and the biosensor. In Kaushal and Wani (2017) study, intelligent fertiliser such as zinc fertilisers with nanoparticles was developed to achieve the controlled release of the fertiliser to the plant roots and sense the feedback while microelectromechanical system (MEMS) was developed to detect the soil quality using microelectronic circuits (Kaushal and Wani 2017).

Moreover, ferric sulphide nanoparticles with agglomeration and sponge like dried structures are produced using green synthesis. The Fourier transform infrared spectroscopy has confirmed the presence of iron and sulphur in the nanomaterial that can enhance agricultural production by using these nano fertilizers (Pavithra et al., 2020).

6.2. Precision Farming

Nanosensors are very small and can detect soil condition, irrigation needs, water and soil pH, nutrient needs, disease and pest occurrence, soil temperature and many important parameters by being distributed over the entire field. Based on the parameters recorded by the nanosensor, appropriate actions will increase crop yields and reduce the unwanted labor resources such as fertilizers, pesticides, etc. This nano sensor concept fits in well with the goals of precision farming or smart farming. Nanosensors, which are made of non-biological materials such as carbon nanotubes, could sense and signal as their small dimensions allow them to act as wireless nanoantenna and collect information from numerous different points (Garcia-Martinez 2016). External devices can then integrate the data to automatically generate incredibly detailed reports and respond to potentially devastating changes in their environment. For example, connected nanosensors monitoring soil or crop condition can automatically alert according to conditions detected by sensors, thus influencing more efficient use of fertilizers, herbicides, pesticides, insecticides, etc. Nanobiosensors are now being developed with all the integrated devices like power source, sensor unit, detector and display unit in a single chip for detecting the plant stress indicators (Kuswandi et al., 2011; Sassolas et al., 2012). Various cues such as increase in sucrose content (Soldatkin et al., 2013), change in concentration of nutrients (Chiou and Lin 2011; Raul et al., 2016) and hormones (Larrieu et al., 2015) etc. can be used for this.

Nanoscale devices are envisaged that would be able to detect and treat disease, nutrient deficiencies or other diseases in crops long before visible symptoms appear. This is the future of farming, an army of nanosensors will be scattered like dust across the farms and fields, working like the eyes, ears and noses of the farming world. These tiny wireless sensors can transmit the information they collect. These are programmed and designed to respond to various parameters such as temperature, humidity and nutrient fluctuations. The distributed intelligence of smart particles can be networked to react immediately to any change in the environment, thus providing advance notice of opportunities and means to deal with environmental variations. Intelligent dust and gas sensors make it possible to assess the number of pollutants in the environment. The most effective approach in this sense is real-time collection of parameters using autonomous sensors linked to a global positioning system (GPS) (Gruere 2012; Hajirostamlo et al., 2015; Mukhopadhyay 2014; Prasad et al., 2014).

6.3. Detection of Pathogens

Plant pathogenic microbes caused diseases in plants and are responsible for huge annual losses that are estimated to be between 13% and 22% globally (Oerke 2006; Wang et al., 2021). However, early detection can be helpful in preventing the yield loss. This requires certain signaling molecules released by pathogens of by host plant in defense response.

Nano-chips like microarray comprised probe which on hybridization produce fluorescence. Hence, Phytopathology, a branch of science which is aims to detect pathogens in agricultural fields. Previously, detection of pathogens was based on observation of disease in plant, which typically developed after several days of infection (Khiyami et al., 2014). Fortunately, due to recent advancement in biosensing technology, it can be detected earlier, and a great loss can be prevented.

6.4. Detection of Biotic Stress

Salmonella spp., *Clostridium* spp., *Vibrio cholera*, *Escherichia coli*, and *Campylobacter* spp. are among the foodborne microorganism pathogens or bacteria that pose a major risk to the animal's production and health and can result in food poisoning, gastroenteritis, etc. Due to contaminated meat and poultry and other sources of foodborne germs, salmonellosis in humans is a worldwide issue (Nowak et al., 2007). There are no analytical tools available to analyse food samples. The quick detection throughout the food chain during storage, transport, and food processing offered by nanobiosensors holds great promise for solving this issue. An electrochemical immunosensor based on liposomal poly (3,4-ethylenedioxythiophene)-coated carbon nanotubes was used to detect the cholera toxin, according to Viswanathan et al., 2006.

Salmonella spp. were found in food by a nanobiosensor mounted antiSalmonella polyclonal antibodies on streptavidin-biotin onto the quantum dot surface, according to Kim et al., (2013). *Salmonella* was found in skim milk by nanobiosensor-based AuNPs (Afonso et al., 2013). The detection of *Escherichia coli* O157:H7 by a nanobiosensor based on functionalized Fe_3O_4 NPs and AuNPs coupled monoclonal antibodies was reported by Wang and Alocijia (2015).

6.5. Detection of Abiotic Stress

Abiotic stress is the harm caused to plants by external, non-living factors such as extreme temperatures, saline water, pollutants, heavy metals, etc. A plant's regular biological processes change as a result of ongoing exposure to these nonliving factors, which eventually results in illness and slows down a plant's development, production, or eventual demise. Therefore, it is crucial to identify illnesses that are caused by abiotic stress in plants. The visual change in the leaves, stem, and roots, thermal analysis, the development of novel receptors and proteins, and the decline in normal functions have typically been used to measure such stress outcomes. The most recent method of creating smart biosensors delivers genetically encoded sensors with the ability of real-time reporting of potential changes in level or activity of the biomolecules of interest.

Vitronectin-like proteins (VN) are found to be significant indicators for tracking plant damage caused by heavy metal stress, such as that caused by Cd(II) and Pb(II), in plants like Arabidopsis and soybean. A quantitative relationship between the biomarker content, the electron-transfer resistance, and the chlorophyll content in plant cells was explained. This phenomenon can be used as a reference technique for early-onset warning systems for heavy metal contamination in the environment, according to the explanation given as invisible damage to plants under the stress of heavy metals (Kumar and Arora, 2020).

Two citrus genotypes were used to extract bacterial extracts or root exudates, and the growth of both rhizobacteria was affected differently depending on how sensitive each citrus genotype was to salt and heat stress. The bacteria that are induced in the root exudates of plants under salt stress were employed as biosensors to gauge plant stress (Kordrostami, Mojtaba et. al. 2021).

Nanoparticles with genetically encoded nanoscale to screen key stress responsive biomolecules. Physiologically related sensitivity with temporal resolutions that may recognize dynamic essential signaling proteins and be recorded and examined on the scale of seconds are now well understood thanks to genetically encoded sensors (Gjetting et al., 2013; Walia et al., 2018). However, using this method in crop plants is constrained by the lack of effective DNA transformation techniques for delivering DNA cassettes or plasmids, as well as by the relatively slow optimization in each crop plant.

By utilising this technique, a foreign plasmid DNA was delivered to the chloroplasts of various plant species, making it feasible to express the yellow fluorescent protein (YFP) without the use of a gene gun. Confocal laser scanning microscopy was then used to assess the expression of YFP. The plastid genome presents a unique opportunity for the development of biosensors that are not restricted to the nuclear genome, considering the tendency of biosensors to localise in the cytoplasm.

For crop plant viral infections, the use of a FRET-based genetically encoded calcium sensor to find and comprehend the underlying signaling network has been thoroughly examined Salicylic acid and ROS, two additional important signaling chemicals, are highly responsive to stimuli and accumulate in chloroplasts. Since their detection and monitoring may be achieved using genetically encoded nano sensors and generalized techniques, they may be a trustworthy indicator of plant health status The previously mentioned innovative recent studies demonstrated progress in creating a platform for an effective phenotyping system using fluorescence imaging (Keinath et al., 2015; Krebs et al., 2012; Loro et al., 2016; Giraldo et al., 2019).

The biological processes such as protein-protein interactions or changes in chemical bonding are the main drivers of genetically encoded nanosensing techniques in plants. In the visible region of the electromagnetic radiation spectrum, these factors affect fluorescence intensity or wavelength. Nietzel et al., (2019) have created biosensors that can detect H_2O_2 in Arabidopsis and have a 15-s temporal resolution that essentially explains the dynamics of H_2O_2 without being sensitive to pH changes during elicitor-induced oxidative bursts.

Numerous research has been successfully used as concept notes, including *in vivo* imaging of reactive oxygen species and redox potential, ATP sensing in tissue gradients and stress, and changed gene expression in plants. Together, these techniques offer up fresh fields for the development of abiotic stress-related plant disease biosensors. Even though there have only been a few research in this field, there are many untapped markets that can be filled by combining nanotechnology with phytopathology-based diagnosis.

6.6. Detection of Pesticides

Chemical substances known as pesticides work to minimise agricultural losses by reducing pest population growth. This increases agricultural output. Pesticides are typically administered in a variety of concentrations to the plants as they grow (Zhao et al., 2015). As a result of its toxicity, the pesticide residue in agricultural products might pose a chemical risk to the food sector. Furthermore, the management of pesticides in the field and the avoidance of pesticide misuse are both facilitated by the detection of pesticide concentration on the plants. As a result, pesticide detection is constantly a crucial part of the hazard analysis essential control points parameter throughout the production of food, and it is receiving study attention from both academic and industrial sides (Sun, D. et. al. 2014). To detect different pesticides, several detection methods have been developed, including biosensors that use surface Plasmon resonance and enzymatic sensors containing acetylcholinesterase that use multiwalled carbon nanotubes and/or single-walled carbon nanotubes.

The most popular method of electrochemical sensors can be used to find different herbicides, insecticides, and pesticides. In order to create one of these nanobiosensors, multiwalled carbon nanotubes and copper oxide nanoparticles were added to the graphite used in hollow fibre pencils. With voltammetry, it was possible to find glyphosate amounts in real-world settings (Mohammed-Bagher et al., 2018).

For the purpose of detecting pesticides at biological interfaces, flexible graphene and terahertz-based sensors with reliable sensing and low cost have been created. Additionally, the commercialization of this pesticide detection gadget will be less expensive because to fewer fabrication steps. For sensing and detection, single step analysis is recommended over multistep methods since it takes much less time. In order to create molecular imprinted polymers that can detect the content of kaempferol in vegetables and analyse their anticancer qualities, a comparable nanosensor has been constructed utilising nanocomposites that contain mesoporous molecular sieves embedded with carbon dots (He et al., 2020).

6.7. Detection of Toxins and Other Pollutants

Many bacterial and fungal infections are linked to the release of toxins, which can have detrimental effects on human health both immediately and over time, such as lowering immunity and changing protein metabolism. To date, piezoelectric, electrochemical, and/or optic sensing techniques have been used to identify the presence of toxins in food products. Heavy metal is just one of the many poisons found in food goods. Heavy metals, such as arsenic, cadmium, and others, can interfere with metabolic pathways and have detrimental effects, as demonstrated by a significant number of clinical research (Bahadır and Sezgintürk, 2017).

Aflatoxin are poisons created by secondary metabolism in fungi that can degrade the quality of foods including rice, almonds, and peanuts. Aflatoxins are known to produce hepatic carcinoma and are thought to be cancercausing. Utilizing the Plasmon resonance phenomenon, a very portable nano biosensor has been created for the detection of these poisons. It was effectively established that a miniaturized test based on spores could detect aflatoxin quickly in milk samples. Despite having all the needed properties, this sensor has several drawbacks that have to do with its ability to be reused, the expense of manufacturing it, and the quantitative analysis of samples. The health effects of eating foods tainted with antibiotic residues have been established, and several foods now fall inside this limit (Moon et al., 2018). Sulfonamide residues in meat and poultry food products must be found. An immunoassay using a nano enzyme label for the detection of sulfadiazine in food residue.

6.8. Detection of Signaling Molecules

Nano biosensors can be used to assess the concentration of signaling molecules made by plants in addition to the uses already described. For instance, the signaling hormone that prevents plant shoot branching was detected using small molecule nano biosensors like strigolactones. Soybeans, pears, and cabbage are just a few of the plants that naturally contain phytoestrogens. They are referred to as dietary estrogens and are produced by plants as part of their defense mechanism against harm from fungus. To date, genistein, daidzein, and resveratrol have all been detected utilizing the FRET probe, which uses fluorescence signals and an estrogen binding domain. due to the crucial functions that dopamine and catecholamine play in the development, growth, and metabolic processes of plants. Nano biosensors were used to measure dopamine levels (Weng et al., 2017).

Similar to this, Yoshimu lactone green was used by Tsuchiya et al., (2015) to create a fluorescence turn-on probe for the detection of strigolactone molecules. Additionally, real-time sensing can be used in conjunction with automation for better resource management. Water is one of the primary natural resources needed for sustainable agriculture in the future, and the automation of irrigation systems employing sensor technology has enormous potential for efficient use of water (Ramnani et al., 2016).

6.9. Integration of Nanotechnology into Biochip Assay Formats

The basis of the diagnosis is on the specificity of the contact of the analyte with the bioassay element (Bakhori et al., 2013; Siddiquee et al., 2014; Walia et al., 2018). Converters used in biosensors include optical, electrochemical, piezoelectric and thermometers. Biosensors can be categorized according to the type of analyte, how the transducer operates and its applications.

Among the range of nano structures used till date for biosensors production, carbon nanotubes showed significant attention (Giraldo et al., 2019). In nano biosensors, reaction takes place which results in alteration of physical/chemical properties which is measured with converter. The alteration results in electronic signals which are generally equivalent to the concentration of analytes.

Conclusion

Biosensors have long been recognised as one of the most potent tools for offering solutions to the world's problems, including climate change, the expanding global population's impact on agriculture, and food safety. By providing ongoing monitoring or early identification of disease outbreaks that can be managed, biosensors can support sustainable agriculture. Biosensors are rapid, efficient, and accurate analytical tools created for the measurement of numerous agricultural sample constituents. Biosensors can therefore satisfy all requirements to expedite the manufacturing of agricultural goods. The incorporation of nanobiosensors in modern agriculture is growing in order to increase productivity in the agricultural and food industries by more effectively utilising natural resources. Nanobiosensors can be used in a variety of applications, like soil assessment, disease management, pathogen identification, adulterant detection, detection of plant stress, and pollutant or heavy metal detection. However, there are very few reports on commercialised nanobiosensors in agriculture. The versatility of the nanobiosensor is another issue that requires more attention. Portable nanobiosensors may become commercially viable if a variety of nanomaterials for biosensing are developed based on bioassays. Farmers will be in a better position to decide on irrigation, fertilisation, insect management, and harvesting using nanobiosensors. A discussion of the near future commercialization of tailor-made nano-sensors in terms of requirements with high specificity and sensitivity can serve as a useful conclusion to the current article.

References

Akshath, U. S., Sagaya S. L., Thakur, M. S. (2012). Detection of formaldehyde in food samples by enhanced chemiluminescence. *Analytical Methods,* 4(3), 699–704. https://doi.org/10.1039/c2ay05608a.

Alexandratos, N. (2009) Highlights and views from MID-2009: paper for the expert meeting on. In: *World food and agriculture to 2030/50*, Rome, 24 June 2009.

Algar, W. R., Krull, U. J. (2008) Quantum dots as donors in fluorescence resonance energy transfer for the bioanalysis of nucleic acids, proteins, and other biological molecules. *Anal. Bioanal. Chem.* 2; 391 1609-1618. https://doi.org/10.1007/s00216-007-1703-3.

Antonacci, F., Arduin, D., Moscone G., Palleschi P., Scognamiglio V. (2017). Nanostructured (bio)sensors for smart agriculture. *TRAC Trend. Anal. Chem.* 98, 95–103.

Bao, J., Huang, T., Wang, Z., Yang, H., Geng, X., Xu, G., Samalo, M., Sakinati, M., Huo, D., Hou, C., (2019). 3D graphene/copper oxide nano-flowers based acetylcholinesterase biosensor for sensitive detection of organophosphate pesticides. Sens. *Actuators B Chem.* 279, 95–101. https://doi.org/10.1016/j.snb.2018.09.118.

Bard, A. (1988) *Electroanalytical Chemistry: A Series of Advances*. Boca Raton, FL, USA: CRC Press.

Bard, A., Parsons R., Jordan J., editors. (1985) *Standard Potentials in Aqueous Solution*. Boca Raton, FL, USA: CRC Press.

Bellingham, B. K. (2011) Proximal soil sensing. *Vadose Zone J* 10:1342–1342. https://doi.org/10.2136/vzj2011.0105br.

Bondar, V. S., Puzyr A. A. Nanodiamonds for biological investigations. (2004) *Physics of the Solid State*;46:761–763.

Boonham, N., Glover, R., Tomlinson, J., Mumford, R. (2008) Exploiting generic platform technologies for the detection and identification of plant pathogens. In *Sustainable*

Disease Management in a European Context; Springer: Berlin, Germany, 2008; 355-363. https://doi.org/10.1007/s10658-008-9284-3.

Budhathoki, N. K., Zander K. K. (2019) Socio-economic impact of and adaptation to extreme heat and cold of farmers in the food bowl of Nepal. *Int J Env Res Pub He* 16:1578. https://doi.org/10.3390/ijerph16091578.

Cao, X., Ye, Y., Liu, S. (2011) Gold nanoparticle-based signal amplification for biosensing. *Anal. Biochem.*; 417:1-16. https://doi.org/10.1016/j.ab.2011.05.027.

Chartuprayoon, N., Rheem, Y., Chen, W., Myung, N. (2010) Detection of plant pathogen using LPNE grown single conducting polymer Nanoribbon. In *Meeting Abstracts*; The Electrochemical Society: Pennington, NJ, USA.

Cheli, F., Pinotti, L., Campagnoli, A., Fusi, E., Rebucci, R., Baldi, A. (2008) Mycotoxin Analysis, Mycotoxin-Producing Fungi Assays and Mycotoxin Toxicity Bioassays in Food Mycotoxin Monitoring and Surveillance. *Ital. J. Food Sci.*; 20: 447-462.

Chesterfield, R. J., Whitfield, J. H., Pouvreau, B., Cao, D., Alexandrov, K., Beveridge, C. A., Vickers, C. (2020) Rational design of novel fluorescet enzyme biosensors for direct detection of strigolactones. *ACS Synth. Biol.* 9 (8), 2107–2118.

Chiou, T. J., Lin, S. I. (2011) Signaling network in sensing phosphate availability in plants. *Annual Review of Plant Biology*. 62; 185-206.

Cui, H. F., Wu, W. W., Li, M. M., Song, X., Lv, Y., Zhang, T. T., (2018a) A highly stable acetylcholinesterase biosensor based on chitosan-TiO2-graphene nanocomposites for detection of organophosphate pesticides. *Biosens. Bioelectron*. 99, 223–229.

Cui, S., Ling, P., Zhu, H., Keener, H. M., (2018b) Plant pest detection using an artificial nose system: a review. *Sensors,* 18 (378).

Dagar, J., Sharma, P., Chaudhari, S., Jat, H. S., Ahmad, S. (2016). *Climate Change vis-a-vis Saline Agriculture: Impact and Adaptation Strategies*. doi: https://doi.org/10.1007/978-81-322-2770-0_2.

Das, A., Das, B. (2019) Nanotechnology a potential tool to mitigate abiotic stress in crop plants. In: *Abiotic and biotic stress in plants. Intech Open*. Available via Dialog. https://www.intechopen.com/books/abiotic-and-biotic-stress-in-plants/nanotechnology-a-potential-tool-to-mitigate-abiotic-stress-in-crop-plants. Accessed 7 Jul 2020.

Deng, H. H., Hong, G. L., Lin, F. L., Liu, A. L., Xia, X. H., & Chen, W. (2016). Colorimetric detection of urea, urease, and urease inhibitor based on the peroxidaselike activity of gold nanoparticles. *Analytica chimica acta*, 915, 74–80. https://doi.org/10.1016/j.aca.2016.02.008

Fernandez-Cruz, M. L., Mansilla, M. L., Tadeo, J. L. (2010) Mycotoxins in fruits and their processed products: Analysis, occurrence and health implications. *J. Adv. Res.,* 1:113-122. https://doi.org/10.1016/j.jare.2010.03.002.

Frasco, M. F., Chaniotakis, N. (2009) Semiconductor quantum dots in chemical sensors and biosensors. *Sensors.* 2009; 9 7266-7286. https://doi.org/10.3390/s90907266.

Fu, Q., Zhang, C., Xie, J., Li, Z., Qu, L., Cai, X., Ouyang, H., Song, Y., Du, D., Lin, Y., Tang, Y. (2019) Ambient light sensor based colorimetric dipstick reader for rapid monitoring organophosphate pesticides on a smart phone. *Anal. Chim. Acta* 1092, 126–131. https://doi.org/10.1016/j.aca.2019.09.059.

Garcia-Martinez, J. (2016) *The Internet of Things Goes Nano,* http://www.scientificamerican.com/article/the-internet-of-things-goes-nano/2016.

Ghassemi, F., Jakeman, A. J., Nix, H. A. (1995) *Salinisation of land and water resources: human causes, extent, management and case studies.* Wallingford (United Kingdom) CAB international.

Gibson, N., Holzwarth, U., Abbas, K., Simonelli, F., Kozempel, J., Cydzik, I., Cotogno, G., Bulgheroni, A., Gilliland, D., Ponti, J., Franchini, F., Marmorato, P., Stamm, H., Kreyling, W., Wenk, A., Semmler-Behnke, M., Buono, S., Maciocco, L., Burgio, N. (2011). Radiolabelling of engineered nanoparticles for *in vitro* and *in vivo* tracing applications using cyclotron accelerators. *Archives of toxicology, 85*(7), 751–773. https://doi.org/10.1007/s00204-011-0701-6 Giraldo, J. P., Wu, H., Newkirk, G. M., Sebastian K. (2019) Nanobiotechnology approaches for engineering smart plant sensors. *Nat. Nanotechnol.* 14, 541–553. https://doi.org/10.1038/s41565-019-0470-6.

Gjetting, S. K., Schulz, A., Fuglsang, A. T. (2013) Perspectives for using genetically encoded fluorescent biosensors in plants. *Front Plant Sci* 4:234.

Gruere, G. P. (2012) Implications of nanotechnology growth in food and agriculture in OECD countries. *Food Policy.,* 37 :191-198. https://doi.org/10.1016/j.foodpol.2012.01.001.

Hajirostamlo, B., Mirsaeedghazi, N., Arefnia, M., Shariati, M. A., Fard E. A. The role of research and development in agriculture and its dependent concepts in agriculture [Short Review]. *Asian J. Appl. Sci. Eng.*

Harz, S., Schimmelpfnning M., Tse Sum Bui B., Marchyk, N., Haupt, K., Feller, K. H. (2011). Fluorescence optical spectrally resolved sensor based on molecularly imprinted determination of maltose, lactose, sucrose and glucose. *Talanta*, 2013; 115: 200-207. https://doi.org/10.1002/elsc.201000222.

He, J., Zhang, L., Xu, L., Kong, F., Xu, Z., (2020) Development of nanozyme labelled biomimetic immunoassay for determination of sulfadiazine residue in foods. *Adv. Polym. Technol.* 7647580, 1–8.

Hosseini, M., Khabbaz, H., Dadmehr, M., Ganjali, M. R., Mohamadnejad, J. (2015) Aptamer-Based Colorimetric and Chemiluminescence Detection of Aflatoxin B1 in Foods Samples. *Acta Chim. Slov*. 62: 721-728.

Hou, W., Zhang, Q., Dong, H., Li, F., Zhang, Y., Guo, Y., Sun, X., (2019) Acetylcholinesterase biosensor modified with ATO/OMC for detecting organophosphorus pesticides. *New J. Chem.* 43, 946–952.

Hussein, H. S., Brasel, J. M. (2001) Toxicity, metabolism, and impact of mycotoxins on humans and animals. *Toxicology,* 167, 101-134.

Iizumi, T., Furuya, J., Shen, Z., Kim, W., Okada, M., Fujimori, S., Hasegawa, T., Nishimori, M. (2017) Responses of crop yield growth to global temperature and socioeconomic changes. *Sci Rep* 7. https://doi.org/10.1038/s41598017-08214-4.

James, A., Franco, F., Merca, E. M., Rodriguez, S., Johny, F., Balidon-Veronica, P., Divina, M., Evangelyn, A., Alocilja, C., Fernando, L. M., (2019) DNA based electrochemical nanobiosensor for the detection of Phytophthora palmivora causing black pod rot in cacao (Theobroma cacao l.) pods. *Physiol. Mol. Plant Pathol.* 107, 14-20. 14–20.

Keerthy, D., John S., Ramachandran T., Bipin G. N., Satheesh Babu T. G. (2014) Pt-CuO nanoparticles decorated reduced graphene oxide for the fabrication of highly sensitive non-enzymatic disposable glucose sensor. *Sens Actuators,* 195:197e205.

Keinath, N. F., Waadt, R., Brugman, R., Schroeder, J. I., Grossmann, G., Schumacher, K., Krebs, M. (2015). Live Cell Imaging with R-GECO1 Sheds Light on flg22- and Chitin-Induced Transient [Ca(2+)]cyt Patterns in Arabidopsis. *Molecular plant*, *8*(8), 1188–1200. https://doi.org/10.1016/j.molp.2015.05.006.

Khiyami, M. A., Almoammar, H., Yasser, M., Awad, M., Mousa, A., Alghuthaymi Kamel, A., Abd-Elsalam, A., (2014) Plant pathogen nanodiagnostic techniques: forthcoming changes? *Biotechnol. Equip*. 28 (5), 775–785.

Krebs, M., Held, K., Binder, A., Hashimoto, K., Den Herder, G., Parniske, M., Kudla, J., Schumacher, K. (2012). FRET-based genetically encoded sensors allow high-resolution live cell imaging of Ca^{2+} dynamics. *The Plant journal : for cell and molecular biology*, 69(1), 181–192. https://doi.org/10.1111/j.1365-313X.2011.04780.x.

Kuila, T., Bose, S., Khanra, P., Mishra, A. K., Kim, N. H., Lee, J. H. (2011) Recent advances in graphene-based biosensors. *Biosens. Bioelectron*. 26 4637-4648.

Kumar, V., Kavita Arora (2020) Trends in nano-inspired biosensors for plants, *Materials Science for Energy Technologies,* Volume 3, Pages 255-273, ISSN 25892991, https://doi.org/10.1016/j.mset.2019.10.004.

Kuswandi, B., Wicaksono, Y., Jayus, A., Abdullah, A., Heng, L. Y., Ahmad, M. (2011) Smart packaging: sensors for monitoring of food quality and safety. *Sens. Instrum. Food Qual. Saf.* 5 (3-4), 137-146.

Larrieu, A., Champion, A., Legrand, J., Lavenus, J., Mast, D., Brunoud, G. (2015) A fluorescent hormone biosensor reveals the dynamics of jasmonate signalling in plants. *Nature Communications*. 6, 1-8.

Li, Y. J., Xie W. H., Fang G. J. (2008). Fluorescence detection techniques for protein kinase assay. *Anal Bioanal Chem*, 390, 2049–57. https://doi.org/10.1007/s00216-008-1986-z.

Li, Z., Xue, N., Ma, H., Cheng, Z., Miao, X. (2018) An ultrasensitive and switch-on platform for aflatoxin B 1 detection in peanut based on the fluorescence quenching of graphene oxide-gold nanocomposites. *Talanta*, 181, 346-351.

Loro, G., Wagner, S., Doccula, F. G., Behera, S., Weinl, S., Kudla, J., Schwarzländer, M., Costa, A., Zottini, M. (2016). Chloroplast-Specific *in Vivo* Ca^{2+} Imaging Using Yellow Cameleon Fluorescent Protein Sensors Reveals Organelle-Autonomous Ca2+ Signatures in the Stroma. *Plant physiology*, *171*(4), 2317–2330. https://doi.org/10.1104/pp.16.00652.

Luan, Y., Chen, J., Xie, G., Li, C., Ping, H., Ma, Z., Lu, A. (2015) Visual and microplate detection of aflatoxin B2 based on NaCl-induced aggregation of aptamer-modified gold nanoparticles. *Microchim. Acta*. 182 995-1001.

Ma, J., Sengupta M. K., Yuan D. K., Dasgupta P. K. (2014). Speciation and detection of arsenic in aqueous samples: A review of recent progress in non-atomic spectrometric methods. *Anal Chim Acta,* 831, 1–23.

Mahmoudi, E., Fakhri, H., Hajian, A., Afkhami, A., Bagheri, H., (2019) Highperformance electrochemical enzyme sensor for organophosphate pesticide detection using modified metal-organic framework sensing platforms. *Bioelectrochemistry* 130, 107348. https://doi.org/10.1016/j.bioelechem.2019.107348.

Mandler, D., Kraus-Ophir, S. (2011) Self-assembled monolayers (SAMs) for electrochemical sensing. *J. Solid State Electrochem.* 15, 1535-1558.

Manjunatha, S. B., Biradar, D. P., Aladakatti, Y. R. (2016) Nanotechnology and its applications in agriculture: A review. *J. Farm Sci.* 29 (1), 1–13.

Mishra, A., Kumar, J., Melo, J. S., Sandaka, B. P., (2021) Progressive development in biosensors for detection of dichlorvos pesticide and review. *J. Environ. Chem. Eng.* 9, 105067. https://doi.org/10.1016/j.jece.2021.105067.

Moon, J. M., Thapliyal, N., Hussain, K. K., Goyal, R. N., Shim, Y. B., (2018). Conducting polymer based electrochemical biosensors for neurotransmitters: A review. *Biosens. Bioelectron.* 15 (102), 540–552.

Mukhopadhyay, S. S. (2014) Nanotechnology in agriculture: prospects and constraints. *Nanotechnol. Sci. Appl.,* 7: 63-71. https://doi.org/10.2147/NSA.S39409.

Negrete, J. C. (2020) Nanotechnology an option in mexican agriculture. J. Biotechnol. Bioinfo. Res. 2 (2), 1–3. https://doi.org/10.47363/JBBR/2020(2)106.

Nietzel, T., Elsässer, M., Ruberti, C., Steinbeck, J., Ugalde, J. M., Fuchs, P., Wagner, S., Ostermann, L., Moseler, A., Lemke, P., Fricker, M. D., Müller-Schüssele, S. J., Moerschbacher, B. M., Costa, A., Meyer, A. J., Schwarzländer, M. (2019). The fluorescent protein sensor roGFP2-Orp1 monitors *in vivo* $H_2 O_2$ and thiol redox integration and elucidates intracellular $H_2 O_2$ dynamics during elicitor-induced oxidative burst in Arabidopsis. *The New phytologist, 221*(3), 1649–1664. https://doi.org/10.1111/nph.15550.

Oerke, E. C. (2006) Crop losses to pests. *J. Agric. Sci.* 144, 31–43. https://doi.org/10.1017/S0021859605005708.

Olkhov, R. V., Parker R., Shaw A. M. (2012). Whole blood screening of antibodies using label-free nanoparticle biophotonic array platform. *Biosens Bioelectron*, 36, 1–5.

Perez-Lopez, B., Merkoçi, A. (2011) Nanoparticles for the development of improved (bio) sensing systems. *Anal. Bioanal. Chem.,* 399 1577-1590.

Prasad, R., Bhattacharyya A., Nguyen Q. D. (2017) Nanotechnology in sustainable agriculture: recent developments, challenges, and perspectives. *Front Microbiol* 8:1014. doi: https://doi.org/10.3389/fmicb.2017.01014.

Prasad, R., Kumar, V., Prasad, K. S. Nanotechnology in sustainable agriculture: present concerns and future polymers and microfluidics. *Eng Life Sci*, 11, 559–65.

Rad, F., Mohsenifar, A., Tabatabaei, M., Safarnejad, M. R., Shahryari, F., Safarpour, H., Foroutan, A., Mardi, M., Davoudi, D., Fotokian, M. (2012) Detection of Candidatus Phytoplasma aurantifolia with a quantum dots fret-based biosensor. *J. Plant Pathol.,* 94: 525-534.

Raul, R. G., Irineo, T. P., Gerardo, G. G. R., Miguel, C. M. L. (2016) Biosensors used for quantification of nitrates in plants. *Journal of Sensors.,* 1-12. https://doi.org/10.1155/2016/1630695.

Rockström, J., Williams, J., Daily, G., Noble, A., Matthews, N., Gordon, L., Wetterstrand, H., DeClerck, F., Shah, M., Steduto, P., Fraiture, C., Hatibu, N., Unver, O., Bird, J., Sibanda, L. Smith, J. (2017). Sustainable intensification of agriculture for human prosperity and global sustainability. *Ambio* 46, 4–17. https://doi.org/10.1007/s13280-016-0793-6.

Safarpour, H., Safarnejad, M. R., Tabatabaei, M., Mohsenifar, A., Rad, F., Basirat, M., Shahryari, F., Hasanzadeh, F. (2012) Development of a quantum dots FRET-based biosensor for efficient detection of Polymyxa betae. *Can. J. Plant Pathol.* 34 :507515.

Sassolas, A., (2012) Prieto-Simon, B., Marty, J. L. Biosensors for pesticide detection: New trends. *American Journal of Analytical Chemistry*, 3:210-232.

Shiddiky, M. J., Torriero, A. A. (2011) Application of ionic liquids in electrochemical sensing systems. *Biosens. Bioelectron.* 26: 1775-1787.

Singh, S., Singh, M., Agrawal, V. V., Kumar, A. (2010) An attempt to develop surface plasmon resonance based immunosensor for Karnal bunt (Tilletia indica) diagnosis based on the experience of nano-gold based lateral flow immuno-dipstick test. *Thin Solid Films.* 519:1156-1159.

Soldatkin, O. O., Peshkova, V. M., Saiapina, O. Y., Kucherenko, I. S., Dudchenko, O. Y., Melnyk, V. G., Development of conductometric biosensor array for simultaneous using cyclotron accelerators. *Arch Toxicol,* 85, 751–73.

Sun, D., Hussain, H., Yi, Z., Siegele, R., Cresswell, T., Kong, L., Cahill, D., (2014) Uptake and cellular distribution, in four plant species, of fluorescently labeled mesoporous silica nanoparticles. *Plant Cell Rep.* 33, 1389–1402.

Thakur, P., Nayyar H. (2013) Facing the cold stress by plants in the changing environment: sensing, signaling, and defending mechanisms. In: Tuteja N., Gill S. S. (eds) *Plant acclimation to environmental stress.* Springer, New York, pp 29–69.

Umasankar, Y., Ramasamy, R. P. (2013) Highly sensitive electrochemical detection of methyl salicylate using electroactive gold nanoparticles. *Analyst.,* 138: 6623-6631.

Walia, A., Waadt R., Jones A. M. (2018) Genetically encoded biosensors in plants: pathways to discovery. *Ann Rev Plant Biol* 69:497–524.

Wang, G., He, X. P., Wang L. L., Gu A. X., Huang Y., Fang B., Geng B. Y., Zhang X. J. (2014) Nonenzymatic electrochemical sensing of glucose. *Microchim. Acta,* 180:161e86.

Wang, Y. P., Pan Z. C., Yang L. N., Burdon J. J., Friberg H., Sui Q. J. and Zhan J. (2021) Optimizing Plant Disease Management in Agricultural Ecosystems Through Rational In-Crop Diversification. *Front. Plant Sci.* 12:767209. doi: https://doi.org/10.3389/fpls.2021.767209.

Xu, Y., Dhaouadi, Y., Stoodley, P., Ren, D., (2020) Sensing the unreachable: challenges and opportunities in biofilm detection. *Curr. Opin. Biotechnol.* 64, 79–84.

Yao, K. S., Li, S. J., Tzeng, K. C., Cheng, T. C., Chang, C. Y., Chiu, C. Y., Liao, C. Y., Hsu, J. J., Lin, Z. P. (2009) Fluorescence silica nanoprobe as a biomarker for rapid detection of plant pathogens. *Adv. Mater. Res.* 79: 513-516.

Zhang, P., Sun, T., Rong, S., Zeng, D., Yu, H., Zhang, Z., Chang, D., Pan, H., (2019) A sensitive amperometric ache-biosensor for organophosphate pesticides detection based on conjugated polymer and Ag-rGO-NH2 nanocomposite. *Bioelectrochemistry* 127, 163–170.

Zhao, C., Liu, B., Piao, S., Wang, X., Lobell, David B., Huang, Y., Huang, M., Yao, Y., Bassu, S., Ciais, P., Durand, J., Elliott, J., Ewert, F., Janssens, I., Li, T., Lin, E., Liu, Q., Martre, P., Müller, C., Peng, S., Peñuelas, J., Ruane, A., Wallach, D., Wang, T.

Wu, D., Liu, Z., Zhu, Y., Zhu, Z., Asseng, S. (2017) Temperature increase reduces global yields of major crops in four independent estimates. *Proc Natl Acad Sci* 114:9326–9331. https://doi.org/10.1073/pnas.1701762114.

Zhao, G., Guo, Y., Sun, X., Wang, X. (2015) A system for pesticide residues detection and agricultural products traceability based on acetylcholinesterase biosensor and internet of things. *Int. J. Electrochem. Sci.* 10, 3387–3399.

Chapter 5

Geno-Sensors: A Future Perspective of Sensing Technologies for Sustainable Development

Maitry Mehta
Anurag Zaveri
Sakshi Sharma
Nirali Vaghani
Ekta Joshi
Avani Zaveri
Dilip Zaveri
and Purvi Zaveri*
Biocare Research (I) Pvt. Ltd., Ahmedabad, India

Abstract

Finding appropriate, accurate tools for measuring changes in physical, chemical and biological parameters around us is a bottleneck for building a transparent monitoring system. Sensors have contributed to the development of technological advancements we observe in sectors like space, security, communication, transport, and domestic services. The development of receptor molecules originating from a biological resource opened a new horizon of opportunity, named as biosensors, in terms of accuracy and precision. In the era where humankind has lost its accountability for the degradation of natural resources, biosensing and monitoring have acted as a torchlight in the middle of the storm. During

* Corresponding Author's Email: purvi250588@gmail.com.

In: Biosensing
Editor: Rushika Patel
ISBN: 979-8-88697-911-4

the COVID pandemic, biosensors proved their presence by detecting the presence of genetic material (genetic sensor/Genosensor), antibodies, antigens and peptides from varieties of matrices to detect the infection and save lives. Genosensors have found a wide range of applications, from contaminant detection to gene sequencing. With the availability of gene editing tools like CRISPR, it has become easier to regulate the expression and overcome lower detection limits of molecule detection by genosensors. This chapter focuses the discussion on genetic material-based sensors and tries to emphasize genosensors contribution in attaining goals towards a sustainable future. The Climate Crisis and cumulative environmental changes we face today remind us how much it will cost to sustain humans. It is important to rightfully invest funds, resources and workforce in hopes of helping mankind grow better and more responsible. Integration of disciplines, inclusivity of research in policymaking and identification of appropriate indicators (like pollutants, genes, species etc.) will pave the path to meeting sustainable development goals declared by United Nations (the Year 2015- 2030).

Keywords: geno-sensor, sustainability, DNA and RNA based genosensors

1. Introduction

Our blue planet has been trapped in crisis due to anthropocentric activities and a consumer-based market approach instead of sustainability-oriented living. A pandemic like COVID-19 has proved that we are far behind in evolving as compared to even the smallest biological entities present in nature. Human health, environment, communities, and survival are at stake. World leaders have started taking climate change seriously, and multilateral agreements are shaping up in global events like the Conference of Parties (COP 27) (https://unfccc.int/cop27) and G20 (https://www.g20.org/) summits. In such a crucial time, where we have a target of 1.5 degrees Celsius and various sustainable development goals by the UN to achieve, we are asking a fundamental question. *Are we prepared and equipped enough to measure change?* What can be detected can only be measured, and what can be measured can only give us insights into trends and current status and guide us for the next steps required. Various indices have been proposed in the last decade to explicitly portray the overall progress in climate change and help us track our actions better. However, they are directly or indirectly based on specific scientifically designed and engineered detection tools to measure the ground reality of health, air, water and soil conditions. This chapter focuses on one such fantastic device, the genosensor and throws some light on the expected roles of the biosensor in sustainable development.

Sensing has been widely used to symbolize the ability to feel or experience something without really being able to communicate in words (concerning human beings). Scientists and engineers have turned the tables around and have gained "expression" to sensing using technology. In an analytical world, sensing can detect any molecule, substance, or object of interest in a particular matrix under predetermined conditions within prescribed limits using various modules. If we try and break each part of the former sentence, it leads us to a multifaceted world of sensors. The sensors have been classified based on various fundamental structural aspects, and there have been various classification systems described elaborately (Wang et al., 2022). For general understanding, broadly, a sensor can be divided based on the following questions (i) *What is it that it detects (analyte)?* E.g., a physical parameter like heat or a chemical entity like pollutant (ii) *How is it that it senses (receptor)?* E.g., by sensing the potentiometric difference in the case of a pH meter or by measuring the quantity of fluorescence generated by organisms (iii) *What is the mode of detection (transducer)*? E.g., digital output like light signals or death/growth of a bacterium. The sensors have also been classified sometimes as generations of sensors with respect to their technological evolution. To understand the application and contribution of genetic sensors, it is essential to visit some structural aspects of the sensor first. The structural parts and possible variations are discussed in detail in the following section.

2. Structural Aspects of Genosensors

A biosensor is a receptor-transducer device with integrated processing power that can turn a biological reaction into an electrical signal. The following figure indicates the essential structural components required to make a functional sensor.

Chemical moieties which are commercially important, as part of the production process or have been identified as a harmful chemical for humans became analytes or substrates of interest in the first place. Thus, the analyte certainly begins the detection process; however, it is not a structural part of the sensor structure. As we can observe, the structure of the sensor consists of the receptor, transducer, electronic components for amplifying detected signals and display units. The types of receptors and mechanisms involved in the first stage of interaction between the sensory part (receptor) and the substrate or analyte determine the type of sensor. It is beyond the scope of this chapter to cover all details of various permutations and combinations available in the market for the development of sensors. However, the type of analyte being detected and key to the sensitivity and specificity of analyte detection lies with the first component of the sensor, i.e., receptor. The transformed analyte can be detected using various transducers like electrochemical (potentiometry, voltammetry etc.), electronic, gravimetric, optical, acoustic, thermal etc. When the receptor molecule of a sensor is originated from a living organism, such sensors are termed Biosensors. There are examples of biomolecules performing the action of the transducer with the help of nanomaterial by researchers in recent years.

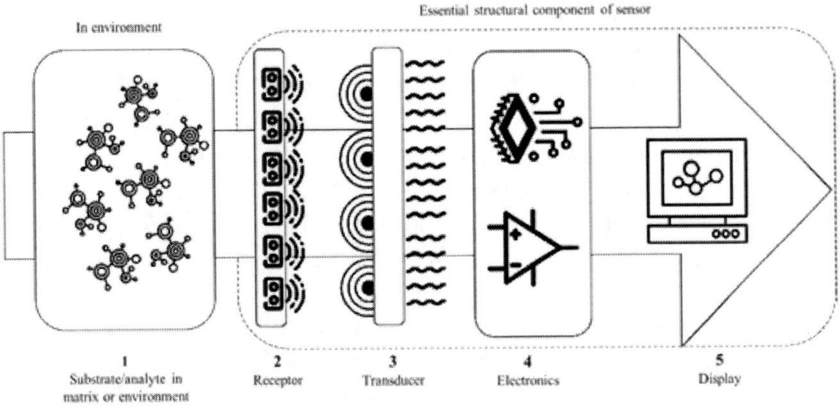

Figure 1. Structural components of biosensor.

Following is the figure indicating major classes of sensor-based mechanisms involved in receiving the signals and passing them on for conversion signals and then to the displaying units.

Physical sensors have found their application in biomedical device development since their commencement. Chemical receptors were a breakthrough as they indicated that any substantial molecular changes could be identified and translated into a detectable signal. With the development of thermal receptors, the industrial revolution received a well-deserved push towards the development of high-end instrumentation facilities. If we look into the details of biological sensors (biosensors), it is evident that the detection happens due to the biological transformation of the analyte or substrate.

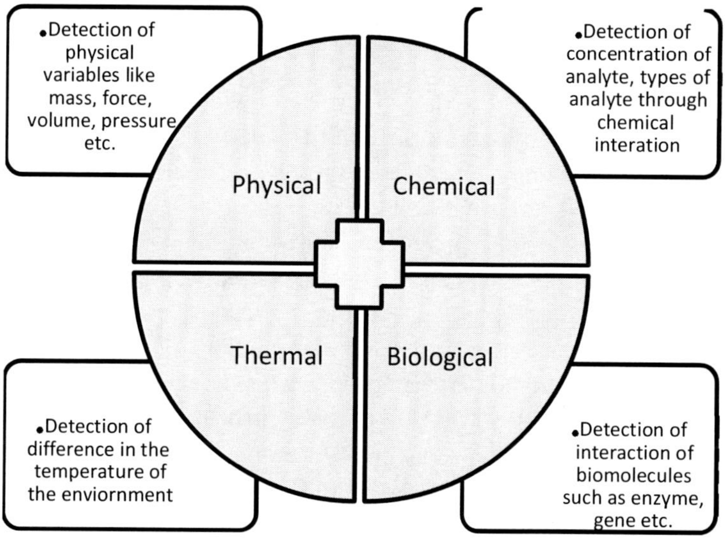

Figure 2. Types of sensors based on the receptor mechanism to convert into signals.

Following are the biological molecules used as receptors by various researchers for the construction of biosensors.

- Enzymes (Guilbault and Lubrano 1973)
- Antibodies (Rudenko et al., 2021)
- Double-stranded Deoxyribonucleic acid (Rasheed et al., 2017)
- Ribonucleic acid (Panel and Grundy 2013)
- Aptamer (single-stranded nucleic acids) (O'Sullivan, 2013)
- Whole cells (Bacteria, yeast etc.) (Simpson et al., 2001)
- Nano material mimicking biomaterial (as receptor and as a transducer) (Handrea-Dragan et al., 2022).

All the biomolecules have their own specificity towards their substrates, and thus they serve as accurate detection systems. In the case of genetic material being used as the receptor, it takes the accuracy and specificity to the level of detection of change in a single nucleotide in the sequence. This characteristic makes DNA and RNA an efficient choice as receptors in biosensor construction.

3. Genosensors as Accurate Successors of Biosensor

The development of biosensors evolved with the development of electrochemistry side by side. Instead of the conceptualization of pH and pH scale in 1909 (Sörensen), for the longest time, any instrument for detection could not come to workbenches due to a lack of understanding of sensory equipment for detecting hydrogen ion concentration. The scientific community was introduced to the biosensing concept by M. Cremer in the year 1906. He demonstrated the concept that the change in acid concentration directly generates electric potential between regions of the fluid present in opposing areas of a liquid (Cremer 1906). Followed by that, Oscar Kjellberg developed electrodes (1907 to 1914), which enabled W.S. Hughes to use an electrode to measure the hydrogen ion concentration in 1922 (Cremer 1906). Meanwhile, between the years 1909 to 1922, enzyme immobilization on aluminium hydroxide and charcoal was successfully done by Griffin and Nelson (Griffin EG and Nelson, 1916). Those techniques contributed indirectly to the development of immobilization techniques of genetic material later on.

The critical discovery of biosensor development was made in 1956 when Leland C. Clark, Jr fabricated the "Clark electrode." The device was initially designed for oxygen concentration detection, and eventually, it was used to detect blood glucose levels by immobilizing the enzyme glucose oxidase on the oxygen electrode (Blair). Leland C. Clark, Jr was entitled as 'Father of Biosensors' for the significant discovery. Currently, many biosensors use a similar concept for detecting an analyte concentration by potentiometrically measuring oxidation-Reduction. 'Yellow Spring Instruments first developed a commercial biosensor in 1975, carving a path for commercial use of the biosensors. Sensing an analyte's concentration based on thermal detection was first used in 1974 by the "Klaus Mosbach group" (Blair). Another milestone was achieved in the evolution of biosensors when optical biosensors (Lübbers and Opitz, 1975) and mediator amperometry biosensors were developed alongside the advancement of other fields. In the historical event of 1953, Watson and Crick illustrated the DNA structure; this discovery opened new possibilities in biosensors. At this time (around 1976), biosensing molecules were widely used in basic analyte-detecting units. Another landmark achievement was the estimation of direct enzyme concentration analysis was made possible by detecting electron movement rather than indirectly determining the potential change in the base fluid (Naresh). With the development of new techniques like DNA-RNA hybridization, it was possible to use DNA as a biosensing molecule. After decades of journey and years of interdisciplinary research, biosensors and nano sensors were built as precision detection tools.

The following section describes the significant techniques and principles used for the development of genetic material-based sensors.

3.1. Use of Double Stranded Deoxyribonucleic Acid (ds DNA) as Receptor

Deoxyribonucleic acid (DNA) is the popular genetic material which stores all the information required for an organism to live. With the development of DNA DNA interactive technologies, it was possible to use this biological substance extensively used for biosensing. DNA meets all the criteria for being a biological receptor due to its exceptional biocompatibility (Morán et al. 2013), thermal stability (Giesen et al., 1998; Morán et al. 2013), and alternate fictionalization (Jäger et al., 2005; Kerman et al., 2013).

DNA-based biosensors consist of DNA strands which can detect the complementary sequence by DNA–DNA hybridization. Deoxyribonucleic acid (DNA) based nano sensors have been developed to play a critical role in vast areas like detecting environmental pollution or monitoring, food analysis, food control, clinical diagnosis, drug discovery and biomedical research (Tichoniuk et al., 2008). DNA biosensors have become the most reliable method in the field of Drug screening and forensic analysis and are able to monitor differential gene expression (Teles and Fonseca, 2008). Not just for detection, DNA-based sensors have eased up a lot of industrial applications and research platforms too. Genetic information gained from genome sequencing has given rise to DNA mass sensing, NGS platforms where nucleotides are used to reveal the complete structure of the target sequence (Wang et al., 2022). This advancement has raised the demand for fast, cheap and high throughput analytical devices to detect molecular diagnosis. Genome sequencing has been able to detect disease causing mutations and human pathogens due to their specific gene sequence.

DNA hybridization biosensors play an impotent role based on complementary DNA sequences. In DNA hybridization biosensors, the detector consists of short single-stranded DNA (ssDNA probe), which is able to form a duplex with complementary target DNA strands with high efficiency and specificity. The ssDNA probe is associated with a transducer which is further able to translate hybridization into a physically measurable value (Palecek et al., 2005).

In this approach, Short single-stranded DNA segments are immobilized on the electrode surface in a way that retains their stability, reactivity, and accessibility. The electrical signal is produced when the target DNA binds to the complementary strand in the process of DNA hybridization (Tichoniuk et al., 2008). Due to the high range of detection targets, a longer lifetime and the cheapest costs of production, the DNA-based biosensors tool has more advantages over conventional biosensors.

The following technologies have been used as mechanisms to prepare DNA-based sensors.

- DNA hybridizations
- Functional DNA Strands-Based Biosensors
- DNA Aptamer Biosensors
- DNAzyme Biosensors
- DNA Templated Biosensors.

3.2. Use of Ribonucleic Acid for Sensor Development

RNA is a versatile molecule that is simply integrated into artificial networks and systems to achieve molecular precision over programmed attributes. Detection of contaminants and toxins and for the insurance of the quality, novel binders for particular targets, i.e., Viruses, toxins, pathogens, proteins, and cell receptors linked to oncology to specify the diagnosis and treatment of diseases, is yet an emerging exploration area. (Ma et al., 2019). As a result of their specificity and ease of accessibility, RNA biosensor has gained much emphasis in cancer diagnosis and prognosis. Currently, nanoparticle-based biosensors based on miRNA are established as possible cancer detection tools (Kaur et al., 2018).

It is essential to recognize various molecular structures and processes also their input and outputs in order to detect various processes. The use of antibodies in such screenings is the conventional method, despite its major cons, such as a pile of costs, longer production time, and poorer thermal stability.

There have been alternative approaches: aptamers are proposed (Catuogno et al., 2016). Aptamers are small, single-stranded DNA or RNA molecules that are chemically produced to act as specific biorecognition components (Kaur et al., 2018). RNA molecules can be used to attain structural and functional diversity. Genetic sequencing of RNA sensors and complementing portable platforms enables the imaging of biomolecular in viable cells (Catuogno et al., 2016). Contamination of water is a considerable economic challenge that causes more than 2 million annual fatalities (Kumar, 2021).

Genosensors have not just evolved in their applications, but with CRISPR-Cas discovery, it is possible to think beyond limits and overtake regulatory mechanisms. This shall bring in the next generation of gene-based sensors. In addition to various detection abilities, nano-sensors also find noble objectives for global goods and community services. The next section attempts to emphasize on current and future impacts of genetic sensors to combat the climate crisis.

4. Sustainable Developmental Goals and Genosensors

The Paris Agreement on 4th November 2016 abided by 194 countries in a legally binding international treaty to help the planet breathe better. It asked governments of various nations to work on reducing emissions, providing climate finance, and limiting the temperature rise to reduce global warming (Agreement, 2015). After that global win for the planet UN developed 17 sustainable development goals and 169 to support actions at various stakeholder levels.

As depicted in the figure, out of the 17 goals declared by the UN, nano sensor contributes directly to the eight most pressing issues. It is very important for an individual to have access to healthy food to maintain good health and well-being. There has been the development of gene-based biosensors have been commercialized, which help in the detection of salmonella from food material. Genosensors help the service provider to serve better quality to the customer and contribute to direct well-being (SDG 3). There have been plenty of studies and research for metal detection from groundwater contaminant detection in wastewater, and indirectly they have been applied in agriculture for pesticide detection in the soil in stressed conditions. The overall impact of detection and monitoring leads to better wastewater treatment and conservation of natural water bodies and resources and leads to the building of better communities with higher resilience (Goal 11). After various multilateral agreements, there has been growing collaboration for technological development and research, where huge capacity building was commenced to support goals 17 and 9, respectively. Also, as all the SDGs have interlinked characteristics, the application of a genosensor for monitoring can have a dominos impact on each of them in a positive manner. As part of an international treaty and commitment to maintain continuous monitoring systems and adhere to transparency clauses, the development of efficient nano sensors and integrating them with data management software shall be the most suitable way.

Genosensors contribution to SDGs

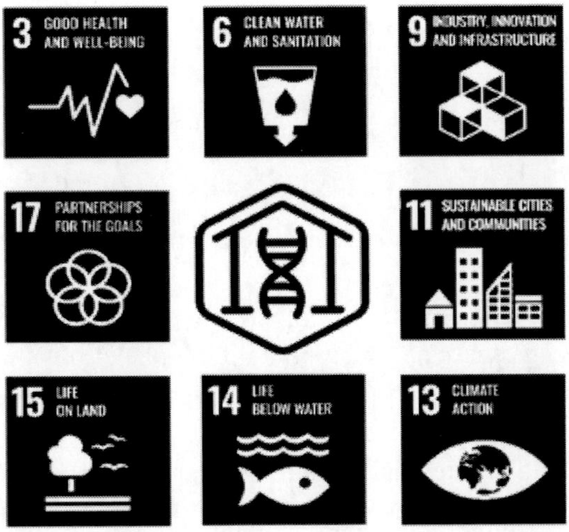

Figure 3. Direct contribution of genetic biosensors towards fulfilment of sustainable development goals.

It would be very inappropriate to believe that such a mighty genetic tool doesn't face any challenges from inception till commercialization. The following section discusses major hiccups a genosensor faces from various angles. These also can be broadly extended to biosensors if the receptor part is exchanged with another molecule of interest.

5. Genosensors and Challenges

Lag period before inception: Due to their delicate molecular structure, it needs years of research to optimize one molecular interaction in housing for it to be converted into a functional genosensor.

- Structural stability: It is difficult to maintain the stable structure of certain intermediate compounds required for detection for long periods of time
- Shelf life: As these are biologically originated molecules, it becomes difficult to maintain the required buffered conditions and temperature before their usage during transportation. Thus, inviting troubles regarding the reliability of the molecules over a period
- Lesser commercialization than research: The genosensor related research has increased since its inception; however, there are only a few commercialized genosensor in the market with ongoing continuous application. This field demands more policy involvement and research support to innovation for generating multiple opportunities for longterm application.
- Real-time monitoring: Genosensors face challenges in implementation for real-time environmental monitoring.

Conclusion

With nucleotide wide accuracy and hybridization efficiencies, genetic material has become choice for development of biological sensors. With Aptamer and CRISPER, the genosensors are expected to provide details of air monitoring and detection of difficult pollutants. With all the contributions expected towards sustainable development goals, it is foreseen that gene-based sensing technologies will be someday implemented in space centres for cultivating habitable situations on other plants too. However, it is important to note that humankind should not take earth for granted and system like genosensors are solutions for sustained growth. It is important that sensing put at right place by policy makers to deliver national commitments Incorporating AI for real time monitoring for efficient system designs can be one of the data management solutions suing integrated sensor development.

References

Agreement P. Paris agreement. In *Report of the Conference of the Parties to the United Nations Framework Convention on Climate Change* (21st Session, 2015: Paris). Retrived December 2015 Dec 12 (Vol. 4, p. 2017). HeinOnline.

Catuogno S, Esposito CL, De Franciscis V. Aptamer-mediated targeted delivery of therapeutics: An update. *Pharmaceuticals*. 2016 Nov 3;9(4):69.

Cremer M. The cause of the electromotor properties of tissue, and a contribution to the science of polyphasic electrolytes. *Zeitschrift Fur Biologie*. 1906 Jan 1(29): 562-608.

Giesen U, Kleider W, Berding C, Geiger A, Ørum H, Nielsen PE. A formula for thermal stability (T m) prediction of PNA/DNA duplexes. *Nucleic acids research*. 1998 Nov 1; 26(21):5004-6.

Griffin EG, Nelson Jm. The Influence of Certain Substances on The Activity Of Invertase. *Journal of the American Chemical Society*. 1916 Mar; 38(3):722-30.

Guilbault GG, Lubrano GJ. An enzyme electrode for the amperometric determination of glucose. *Analytica chimica acta*. 1973 May 1; 64(3):439-55.

Handrea-Dragan IM, Botiz I, Tatar AS, Boca S. Patterning at the micro/nano-scale: Polymeric scaffolds for medical diagnostic and cell-surface interaction applications. *Colloids and Surfaces B: Biointerfaces*. 2022 Jul 26:112730.

Jäger S, Rasched G, Kornreich-Leshem H, Engeser M, Thum O, Famulok M. A versatile toolbox for variable DNA functionalization at high density. *Journal of the American Chemical Society*. 2005 Nov 2; 127(43):15071-82.

Kaur H, Bruno JG, Kumar A, Sharma TK. Aptamers in the therapeutics and diagnostics pipelines. *Theranostics*. 2018;8(15):4016.

Kerman K, Kobayashi M, Tamiya E. Recent trends in electrochemical DNA biosensor technology. *Measurement Science and Technology*. 2003 Dec 11; 15(2):R1.

Kumar P. Climate change and cities: challenges ahead. *Frontiers in Sustainable Cities*. 2021 Feb 25; 3:645613.

Li M, Zhou X, Guo S, Wu N. Detection of lead (II) with a "turn-on" fluorescent biosensor based on energy transfer from CdSe/ZnS quantum dots to graphene oxide. *Biosensors and Bioelectronics*. 2013 May 15; 43:69-74.

Lübbers DW, Opitz N. The pCO2-/pO2-optode: a new probe for measurement of pCO2 or pO in fluids and gases (authors transl). Zeitschrift fur Naturforschung. *Section C, Biosciences*. 1975; 30(4):532-3.

Ma Y, Geng F, Wang Y, Xu M, Shao C, Qu P, Zhang Y, Ye B. Novel strategy to improve the sensing performances of split ATP aptamer based fluorescent indicator displacement assay through enhanced molecular recognition. *Biosensors and Bioelectronics*. 2019 Jun 1; 134:36-41

Morán, MC, Nogueira, DR, Vinardell, MP, Miguel, MG, Lindman, B. Mixed protein–DNA gel particles for DNA delivery: Role of protein composition and preparation method on biocompatibility. *Int. J. Pharm*. 2013, 454, 192–203.

O'Sullivan CK. Aptasensors–the future of biosensing?. *Analytical and bioanalytical chemistry*. 2002 Jan; 372(1):44-8.

Palecek E, Scheller F, Wang J, editors. *Electrochemistry of nucleic acids and proteins: towards electrochemical sensors for genomics and proteomics*. Elsevier; 2005 Dec 19.

Panel ED, Grundy SM. An International Atherosclerosis Society Position Paper: global recommendations for the management of dyslipidemia. *Journal of clinical lipidology*. 2013 Nov 1; 7(6):561-5.

Rasheed PA, Sandhyarani N. Electrochemical DNA sensors based on the use of gold nanoparticles: a review on recent developments. *Microchimica Acta*. 2017 Apr; 184(4):981-1000.

Rudenko N, Fursova K, Shepelyakovskaya A, Karatovskaya A, Brovko F. Antibodies as Biosensors' Key Components: State-of-the Art in Russia 2020–2021. *Sensors*. 2021 Nov 16; 21(22):7614.

Simpson ML, Sayler GS, Fleming JT, Applegate B. Whole-cell biocomputing. *Trends in biotechnology*. 2001 Aug 1; 19(8):317-23.

Teles FR, Fonseca LP. Trends in DNA biosensors. *Talanta*. 2008 Dec 15; 77(2):606-23.

Tichoniuk M, Ligaj M, Filipiak M. Application of DNA hybridization biosensor as a screening method for the detection of genetically modified food components. *Sensors*. 2008 Mar 27; 8(4):2118-35.

Wang Q, Wang J, Huang Y, Du Y, Zhang Y, Cui Y, Kong DM. Development of the DNA-based biosensors for high performance in detection of molecular biomarkers: More rapid, sensitive, and universal. *Biosensors and Bioelectronics*. 2022 Feb 1; 197:113739.

Index

About the Editor

Dr. Rushika Patel
Research Associate, Gujarat Biotechnology Research Centre,
Department of Science and Technology,
Government of Gujarat, India and
Governing Council member of Wildlife & Conservation Biology
Research Foundation, India
Email: rush2907@gmail.com

Dr. Rushika Patel is currently associated as a Research Associate with Gujarat Biotechnology Research Centre, DST, Government of Gujarat. She is working on a project entitled "Evaluating the Success of PanchKarma, an Ancient Ayurvedic Treatment in Rheumatoid Arthritis Through Biotechnology". Apply multi-omic approach to analyze gut microbiome, host-genetic factor and metabolome under Rheumatoid Arthritis condition and during treatment. Her project work received the best paper award in the GTU-ICON 2022 international conference.

She has done her PhD from Nirma University and developed "The fluorescent protein based biosensing strains for Aromatic Hydrocarbon detection in aqueous system". Her work also received the best poster award in ICMR 2018. She also received numerous awards in biodiversity conservation and waste management. She was invited as resource person in the various program i.e. Faculty Development Program, Workshop, hand-on-training program and Webinars. She was reviewer of BIRAC-SITARE-GYTI awards 2020, Government of India. She is a coordinator of Scientific and Academic services provided by Wildlife and Conservation Biology Research Foundation.